Praise fo
Made from S

"This fine, simple book is the real deal — and it w… relief to people feeling some silent dread in a time of rising gas prices, food shortages, and the like. Much can be done — in your home!"
— **Bill McKibben,** author of *Deep Economy*

"This is an outstanding book for anyone yearning for the satisfactions that come with a simpler, more self-reliant, and sustainable life. I highly recommend it, for both country and city homesteaders."
— **Cheryl Long,** Editor in Chief, *Mother Earth News*

"If you're tired of being just another consumer, and want to take charge of creating your own life, this book is for you. It has both the how-to and the why-to. It reads like fiction but delivers a wealth of useful, down-to-earth information."
— **David Wann,** author of *Simple Prosperity* and coauthor of *Affluenza*

"Woginrich writes with an infectious enthusiasm and a dry wit that may have you ordering hens before you reach the last page. A delightful introduction to the simple (and not-so-simple) life."
— **William Alexander,** author of *The $64 Tomato*

"I can't get enough of Woginrich's life on her Vermont farm . . . this book left me wanting much, much more."
— **Debbie Stoller,** *Bust*

"It's a how-to as well as a what-not-to-do."
— ***Boston Sunday Globe,*** "Shelf Life"

"This book isn't about having a farmhouse on acres of land, or a barn full of livestock, but about being more open to learning the simple skills most of us have forgotten."
— ***Deseret News***

"*Made from Scratch* is about being more open to learning the simple skills most of us have forgotten, and finding joy in the process."
— **Homegrown.org**

"The book is chockablock full of 'simple life' advice on everything from creating storage from scratch to gardening, with loads of 21st-century homespun philosophy to boot."
— ***Milwaukee Journal Sentinel***

"Her essays, supplemented with how-tos, are philosophical, humorous, and remarkably poised for a newbie writer."
— ***Seattle Post-Intelligencer***

"Woginrich's comfy writing style and gentle humor make this book a must-read for anyone who dreams of a simpler, handmade life."
— ***ForeWord,*** November 2010

The mission of Storey Publishing is to serve our customers by publishing practical information that encourages personal independence in harmony with the environment.

Edited by Carleen Madigan
Art direction and text design by Mary Winkelman Velgos
Cover design by Dan O. Williams
Text production by Liseann Karandisecky

Front cover illustration by © Meg Hunt/Scott Hull Associates
Author photograph by © Tim Bronson

Storey Publishing
210 MASS MoCA Way
North Adams, MA 01247
www.storey.com

Printed in the United States by Versa Press
10 9 8 7 6 5 4 3 2 1

Library of Congress Cataloging-in-Publication Data

Woginrich, Jenna.
Barnheart / by Jenna Woginrich.
p. cm.
Includes index.
ISBN 978-1-60342-795-1 (pbk. : alk. paper)
1. Country life—Vermont. I. Title.
S521.5.V5W64 2012
630.9743—dc23

2011024846

THE INCURABLE LONGING FOR A FARM OF ONE'S OWN

a memoir

Jenna Woginrich

Storey Publishing

Thanks So Very Much

Thank you to my parents, Pat and Jack, who have watched with grace and support (even when it confused the hell out of them) as their daughter evolved from an urban designer to a rural shepherd. My parents are the only reason I ever believed I could achieve whatever I wanted in this world, even if what I wanted was to be sitting with a flock of sheep on a hill.

Thank you to my brother and sister, John and Kate, who relentlessly support me, nod in approval, and make me laugh at my own antics. Thanks to Kevin Boyle, who has been my best friend for over a decade and has always loved me, even when I let him down (which I do from time to time). Thank you to Erin Griffiths, Raven Pray, Leif Fairfield, Phil Monahan, Steve Hemkens, Sara and Tim Mack, and Nisaa Askia — all of you are part of this story. It's the folks who stay in touch, visit, call, and heckle that make a life that makes a book.

And of course, big thanks to everyone at Storey, most of all to my grand editor, Carleen Madigan, whose own work and writing fuel my dreams and constantly make me want to learn new skills and take on new adventures. Thank you to everyone at Orvis — coworkers and friends who are wonderfully tolerant of me (possibly the most unorganized person in New England to hold down a day job) and make that place a dream career. Not a lot of offices let you bring your kid to work (when it's a bottle-fed Toggenberg).

Thank you to James Daley and Phil Bibens, who have helped me build sheep sheds and amazing friendships. You guys made Vermont feel like home first. Thanks to everyone at Wayside, especially Doug and Nancy Tschorn, who are my unofficial grandparents and run the best country store in the state of Vermont. This fact can be argued, but the argument is pointless. Thank you to Tim Bronson, who helped me get this book started and earnestly supported me at a point when the book you're holding in your hands was just a shot in the dark in a conference room. Thank you, Paul Fersen, fellow Civil War buff, farmer, and friend who keeps me laughing and lends me old calf hutches he's not using to house my bum goats. Thank you to Eric and Erica Weisledder, Suzanne and Allan Tschorn, Jo and Bob Wise, Nancy and Dean Bishop, the Daughton family and Roy next door — neighbors like you made this farm (and the future of this farm) possible. Thank you, members of NEBCA, especially Barb and Denise, important members in the great club of shepherds who are just beginning to show me the ropes of this new life. Thank you all, over and over.

And thanks, of course, to Jazz and Annie, still the best roommates a girl could ever have, and Gibson, the finest farmhand and business partner I've ever known.

For Mom and Dad

The only reason everything happened

How to Tell If You're Infected

Certain people, myself included, are afflicted by a condition that's difficult to describe. It's not recognized by physicians or psychoanalysts (yet), but it's really only a matter of time before it's a household diagnosis. It's a sharp, targeted depression, a sudden overcast feeling that hits you while you're at work or standing in the grocery-store checkout line. It's a dreamer's disease, a mix of hope, determination, and grit. It attacks those of us who wish to God we were outside with our flocks, feed bags, or harnesses instead of sitting in front of a computer screen. When a severe attack hits, it's all you can do to sit still. The room gets smaller, your mind wanders, and you are overcome with the desire to be tagging cattle ears or feeding pigs. (People at the office water cooler will stare and slowly back away if you say this out loud. If this happens to you, just segue into sports banter and you'll be fine.)

The symptoms are mild at first. You start reading online homesteading forums and shopping at cheese-making supply sites on your lunch break. You go home after work and instead of turning on the television, you bake a pie and study chicken-coop building plans. Then somehow, somewhere along the way you realize that you're happiest when you're weeding the garden or collecting eggs from the henhouse. It's all downhill from there. When you accept that a fulfilling life requires tractor attachments and a septic system, it's too late. You've already been infected with the disease.

This condition is roughly defined as the state of knowing unequivocally that you want to be a farmer but, due to personal circumstances, cannot be one just yet. So there you are, heartsick and confused in the passing lane, wondering why you can't stop thinking about heritage-breed livestock and electric fences. Do not be afraid. You are not alone. You have what I have. You are suffering from Barnheart.

But do not panic, my dear friends; there is a remedy! The condition must be fought with direct, intentional actions that yield tangible, farm-related results. If you find yourself overcome with the longings of Barnheart, simply step outside, get some fresh air, and breathe. Go back to your desk and finish your office work, knowing that tonight you'll be taking notes on spring garden plans and perusing seed catalogs. Usually, those small, simple actions that lead you in the direction of your own farm can help ease the longing.

At times, though, you might find yourself resorting to extreme measures — calling in "sick" to work in the garden, muck out chicken coops, collect eggs, and bake bread. After all, this is a disease of inaction, and it hits us hardest when we are furthest from our dreams. If you find yourself suffering, make plans to visit an orchard, a dairy farm, or a livestock auction. Go pick berries at a local U-pick farm. Busy hands will get you on the mend.

And when you find yourself sitting in your office, classroom, or café and your mind wanders to dreams of the farming life, know that you are not alone. There are those of us who also long for the bitter scent of manure and sweet odor of hay in the air, to feel the sun on our bare arms. (I can just about feel it, too, even in January, in a cubicle on the third floor of an office building.) Even though we straighten up in our ergonomic desk chairs, we'd rather be stretched out in the bed of a pickup truck, drinking in the stars on a crisp fall night.

When your mind wanders like this and your heart feels heavy, do not lose the faith, and do not fret about your current circumstances. Everything changes. If you need to stand in the slanting light of an old barn to lift your spirits, go for it. Perhaps someday you'll do this every day. For some, this is surely the only cure. I may be such a case.

We'll get there. In the meantime, let us just take comfort in knowing we're not alone. And maybe take turns standing up and admitting we have a problem.

Hello. My name is Jenna. And I have Barnheart.

A CABIN IN THE WOODS

I HAVE A BAD HABIT OF RENTING HOUSES from the other side of the country. When I was moving to Idaho, I found an old farmhouse online and made arrangements to live there from an apartment in Knoxville, Tennessee. There was no reconnaissance trip out West to find a home, kick floorboards, and inspect chimneys. This had to be quick and dirty: a lease, a check, a handshake over telephone wires.

When I accepted my current job in Vermont, I needed to do that dance again. Most people with a bit of cash and some luck can find a small apartment and waltz right into their new town. I, however, was developing this agricultural habit and needed a rental that could handle my desire to turn someone else's backyard into a farm. The hunt was on.

I called a half-dozen places before I started losing heart. Landlords are thrilled to hear you're a young, responsible professional, imported to work at a big-name company, but when you start asking if they've ever had their soil tested and what their stance on roosters is, they suddenly have a cousin who really needs the place. Getting a rental house isn't an easy thing to do when you want to rip up sod, install a coop of chickens, and share your living quarters with two large, shedding dogs. Yet I asked, and searched, and begged. A girl's got to try.

I was on the prowl for a new home somewhere in Bennington County, which was close to work and seemed to have a lot of farms around it. I scoured the Internet and local Craigslist ads. I looked up local papers and pennysavers online. My sanctuary had to be out there somewhere.

I had been able to get a taste of the place when I was flown in for an interview a few weeks before my search got into the teeth-grinding stage. I drove around back roads and land outside Manchester and was unsettled at the price of rentals I came across. Some places cost more to rent for a week than I'd be making in a month. Then I remembered the newspaper I'd folded up and stashed in my suitcase during my interview the week before.

I knew nothing specific about the geography of the place — only the address of the office — and I was hoping to stake my claim somewhere within a twenty-minute drive. An ad in my crumpled cross-continental *Manchester Journal* had a listing that caught my attention. I read it with raised eyebrows, my dogs curled at my feet under the kitchen counter. A cabin was available in a town called Sandgate. It came with six acres of pasture and woods, two streams, running water, electricity, a fireplace, and an oil furnace. It had a bedroom, kitchen, living room, and half bath. I did a little homework and discovered it was only eleven miles from my new desk. This was too good to be true.

I called the landlord right away. Either out of desperation to fill the rental or apathy about her property's lawn, she agreed to rent me the six-hundred-square-foot cabin and surrounding land. She approved the dogs, seemed fine with the idea of a garden (there was even a gated garden on site), and didn't scream at the notion of chickens. I sent her a check and made arrangements to meet a neighbor to hand me the key on move-in day. My whole body relaxed when that lease was signed. I had no fears about driving cross-country or starting a new job, but the idea of moving to a brand-new place without my own bedroom waiting for me was terrifying. Now, not only did I have a destination, I had a home as well. I was off to find a cabin in the woods twenty-eight hundred miles away.

The drive east was surreal and miserable. I made my way through a blizzard on a Montana mountain pass, stayed in the worst hotel in recorded history, and fought a serious case of the flu, which caused me to pull over and heave an impolite offering along the battlefield trail at Little Big Horn. Jazz and Annie watched me from the back windows of the car. It was a cold, clear day in Montana, and the view was stunning. I panted from the flu. My dogs panted inside the wagon from wanting to chase down the wild ponies foraging among the dead trees. I was suspicious of my condition. The nerves about moving to a new place were likely responsible for my stomach. Spending a year in a small town, making close friends, discovering a whole new way of life — these things don't make for an easy transition, especially when I'd never intended any transition in the first place.

I hadn't left Sandpoint by choice; I'd been forced out by fate and circumstance. I lost my job in a company-wide firestorm of layoffs, and with the way the economy was taking a nosedive, I was just grateful to have found new work. But gratitude is not the remedy for anxiety. I shrugged off whatever doubt I had left and got back in the car. I had days of travel ahead of me, and I wanted to go home. The sad part was that I had no idea what "home" meant anymore.

I would be building a whole new life in Vermont. I wanted it to somehow resemble my previous one in Idaho. Out there I had dedicated myself to learning basic country skills and starting a small homestead. My rented farm was home not only to me, but also to my first-ever organic vegetable garden, hive of honeybees, flock of prize-winning chickens, and much more. It was in Idaho that I learned to sew, knit my own wool hats, and bake my first loaf of homemade bread. I'd fallen in love with homesteading and wanted to continue falling in Vermont. Hell, maybe even do more than I did out West? I now had six acres to work with. A secret part of me dreamed of adding to my guest list. I desperately wanted sheep, goats, a border collie, and a pickup truck. I wanted to double my gardens, sew my own cowboy shirt, and wear my beat-up old hat into a Vermont feed store and have people know me by my first name. Gary Snyder, one of my heroes, famously said, "Find your place in the world and dig in." I thought, "I'm ready, Gary. Just hand me a friggin' shovel."

It's good to want things. Or at least that was my optimistic motto. I had no concrete idea what the property would be like or what opportunities the new address would grant me. I knew about the garden, but in photos it was covered in snow; who knew what lay under the crust of ice? I could make do with gardening in containers, but the idea of renting six acres and not putting a hoe to them felt horribly wrong. I also wanted to get chickens in the spring and was secretly terrified my landlord would limit the livestock to my two dogs, which she graciously allowed in a house smaller than my parents' living room. Of course, I dared not raise the idea of bees or rabbits. I was focused on just getting there and starting a new job in a new state, a state I knew nothing about, except for Burton snowboards and Ben & Jerry's ice cream. I knew it was a farm-friendly, green-leaning, open-minded place, so that was working in my favor. The idea of an absentee landlord who would allow me to grow heirloom tomatoes and raise a few chickens seemed totally plausible. With that hope in mind, I kept driving east.

After what felt like an eternity of days run on raw nerves and DayQuil, I arrived in the land of Green Mountains and black-and-white cows. It was mid-February when we crossed the final state line from New York into Vermont. Bennington County was coated in ice. There were no picturesque snow-covered barns or white-tipped branches. Just ice. Angry, tenacious, road-skidding ice. As we rolled away from the town of Arlington, up into the Taconic Mountains, we passed a sign that read **SANDGATE EST. 1761.**

Sandgate wasn't so much a town as it was a village. And even calling it a village was a bold overassessment — there was one whitewashed town hall and a Methodist church. A collection of farmhouses and livestock scattered around eleven square miles. All the houses we passed in the afternoon sunset had smoke coming out of their chimneys, lights on inside, and a few horses by yard fences. But aside from this scattered evidence of human life, the place seemed empty. No one was outside. I wasn't expecting a welcoming parade or anything, but it did come across a little colder than this new kid in town preferred. I guess the locals had no need to run to their windows and see the ten billionth Subaru roll past their houses.

I turned the last right at the end of a series of dirt roads, and there she was: tucked under a pair of giant pine trees and squarely situated on top of a hillside was my new cabin. I parked the car and nearly ran up onto the porch, loving the high, red roof and thinking, as I turned the corner to the screen door, "My God . . . this is all mine! For at least a year, this is all mine. . . ." I grinned like an idiot.

Then I heard a voice calling my name. It was the neighbor. She seemed to be in her early sixties, with long gray hair, a pair of heavy carpenter pants, and a wool sweater. After handshakes and a brief chat about the poor road conditions, she showed me around. We walked around the perimeter of the small home that already felt like the HQ of my future homesteading empire. She talked about important things like septic tanks and oil deliveries, but I could only think about where the chickens' brooder box should go. Finally, she handed over the key. I waved to her as she walked down the hill to her property, just a hundred yards away. I liked that someone was close. As she disappeared through the pines to her own cabin, I turned to the big green door in front of me. An old-fashioned doorbell that you turn like a key jangle-rang as I cranked it. I laughed at the simple design; the sound was friendly. After a good look up and down, I unlocked the door and went inside.

Inside was a perfect little kitchen with a cork floor and a few conventional appliances that seemed fairly new. It was the biggest room in the cabin, and I could walk across it in six steps. Off to its right was a small living room with a fireplace and a wooden futon. My inner design student tends to snub futons, but this one seemed to have some substance. On top of it were some blankets and a quilt. The room shone with some sparse furnishings of lamps, end tables, and an entertainment center, which I would turn into my bookcase (I don't have a TV). On the kitchen table was a stack of local maps, menus, and contact information. Off to the left was a bedroom without a bed. The movers wouldn't be here for a few days, so I was silently grateful for the futon. I'd eat some interior-decorating crow. It beat sleeping like one.

That first night in our new home, it was just the dogs and me. Whoever had prepped the cabin had left us some dry wood, so I started a roaring fire. While it heated up the small house, I made a warm bed out of the blankets and unpacked my Sherpa of a station wagon. Inside the packed car were all sorts of things I just didn't trust a moving van to deliver intact. I'd brought along with me my most prized possessions, mostly musical instruments. My fiddle, dulcimer, and banjo were along for the ride, and so were my favorite antiques — a ceramic dinosaur, a music box with a mechanical dancing President

Nixon, and an ancient Lassie stuffed animal with a plastic face. I unpacked a Fire King jadeite mug and a stove-top percolator. (Coffee is something I don't go four hours without unless I'm unconscious.) I set up my computer on a coffee table against the cabin wall, so I could watch a movie before bedtime.

Between the fire, the warm dogs on both sides of me, and the fiddle propped up on one of the cabin's wooden chairs, I relaxed completely. I had made it. Through a blizzard, the flu, and miles of icy roads, I had made it to this cabin at the end of the world. I didn't know a single person. Hell, I didn't even know where to buy breakfast cereal, but for tonight I was all set. It didn't feel like home yet, but it did feel like the beginning of one.

The next morning I woke up a Vermonter. (I know locals around here cringe when transplants call themselves that, but as far as my income-tax files were concerned, I was now a resident of the Green Mountain State.) I made coffee, took the dogs out for a walk, and got my first real look at the place. To my surprise, the garden looked fantastic — a large twenty-by-fifty-foot plot with a wire fence, gate, the works! It wasn't exactly in great shape, though. It looked like no one had tended it in years, but the possibilities were enough to cause a big smile to slide across my face. The ground that wasn't covered in patches of ice showed dead grass, even and untilled. The gate was apparently being held together by spite and tetanus. The fence posts were falling down. It had been easily a decade since this soil had hosted a pumpkin on a vine or a row of sweet corn. But the potential was like a shot of adrenaline. The first real sign of homestead life was beyond that garden gate. There would be food.

Behind the old garden was a metal shed. It was a small structure, the kind of place where people park a riding mower and some rakes. I sneaked inside, past the broken door, and discovered nothing but old flowerpots and cookie tins inside. The walls seemed solid, though, and the roof just as sound. It had a dirt floor and some old pieces of a long-ago-orphaned carpentry project. The place obviously hadn't been used since the garden was in production. All I could think about was chickens.

This old shed would be a veritable hen mansion. In Idaho I had ten birds living in two tiny coops; this shed could easily host twenty layers and a proud rooster. Maybe even a pair of geese or a few ducks. My eyes scanned the property, my mind gathering ideas. The open area around the cabin seemed to be about an acre, maybe slightly more. The clearing was surrounded by a windbreak of trees. I could hear the rushing of the cold creek that circled the property line. There was a rat's nest of field fencing behind the shed, possibly used to hold leaves for composting at the edge of the woods. A few cinder blocks were stacked by an old woodpile. I could already imagine a hive of bees swarming there, happily buzzing, their legs heavy with yellow pollen.

I sharply inhaled a lungful of cold air, and Jazz looked up at me as if something were wrong. "Good Christ . . ." I realized aloud, "this place is going to change everything."

This was hope, folks. This was exactly the place I needed to continue my homesteading aspirations. I could build and expand on everything I had learned in Idaho, maybe even start setting down roots. This place was primed and ready for someone who wanted to really use it. I already had the collateral of a year's experience and the will to work as hard as I needed to. It felt like fate herself had landed me at this little cabin with a garden, a coop, and even some pasture waiting for me. Beyond the garden was a clearing about half an acre in size. I looked out into the dead tall grass and let myself close my eyes, seeing it in my mind as high summer with sheep munching away. The likelihood of sheep on a rented parcel was about as realistic as the landlord's letting me turn her garage into a high-stakes casino. But a girl can dream. In a New England winter, dreams keep you warm.

My heart was pounding. My eyes were tearing up. I knelt on the ground to hug Jazz and Annie, who were sitting alongside me. "Guys," I whispered to them, "this is going to be wonderful. Just wait and see." Jazz and Annie didn't comment. Siberian huskies are known for being professionally stoic as they age, but they got it. Both looked up at me with wagging tails and panting behind wolfish smiles. They didn't know it yet, but the small dirt roads of Sandgate were perfect for their small kick-sled. We would be able to mush for miles here and not worry about running into highways or herds of livestock like we did in Idaho. Or maybe they did know? They smelled something good in the air, their black noses lifting to the rush of pine needles and smoke from the neighbor's woodstove. I wouldn't put it past Jazz. He understands everything.

If luck could cripple, I was limping. The property seemed to have endless possibilities. It could host a garden three times the size of my raised beds in Idaho and produce enough vegetables, eggs, honey, and angora wool to keep me stocked and occupied all through the next winter. I had no idea how long the cabin could be mine, but I was certain of the next twelve months. As long as I kept paying rent, this place would in turn earn its keep for me by providing me with most of my food and entertainment. I'd have a full growing season ahead of me. By the time the last of my Cherokee Purple tomatoes were dropping off the vines, I would be welcoming my first Vermont autumn. It is staggering how much I looked forward to October. November 1 is my least favorite day of the year.

This year the high harvest would include more than just pretty leaves and red-covered bridges — to me it would be a round of applause. Even though the place was a barren tundra at the moment, there would be blood pumping in its veins in a few short months. Young pullets would scatter themselves through the green grass. Peas and squash would swirl around that old dead fence and make it come alive again. Visions of gray geese on that green-painted porch, next to a hutch of Angora rabbits began to inhabit my already overstimulated mind. This property was going to take all the lessons I'd learned in Idaho and turn them into advanced courses. Now I just had to get to work. There was a lot of planning, a slew of phone calls, and some amount of begging for permission still ahead of me. But in my gut I already knew that this year was going to be okay. It had to be.

A VERY LONG WINTER

IT FELT LIKE WINTER WOULD NEVER LEAVE. All of southern Vermont was filled with grimy snow, turned gray from mud and exhaust. Driving to work was like driving through a war zone of sludge, naked trees, icy turns, and barren fields. This was Mud Season. Part of me was happy I'd moved here in the most climatically desperate (not to mention least touristy) time of the year, because I'd really appreciate spring when it finally did arrive. But the other part of me just thought the place looked beat, dead, and unlivable. I had all these big plans for poultry and a garden, and right now the place reminded me more of Chernobyl than the scenes from *The Sound of Music* that the Vermont board of tourism brochures had promised me.

Weekends of exploring grocery stores, movie theaters, and Laundromats took over the borrowed time I now had in my schedule. All my homesteading plans were on hold until the thaw came. I couldn't slam a hoe into the ground or bring home chicks until late April or early May. Northern New England is a place that takes winter just as seriously as the Pacific Northwest does, and that year the cold season seemed to last lifetimes longer than the winters of my past. It had started before I'd left Idaho, with the first snow in October. The succeeding weeks pounded Sandpoint with so much powder that by the time I left in February, the driveway had plow piles close to ten feet high, and even the flattest, driftless areas were topped with four and a half solid feet of snow. (You know you're dealing with a different kind of winter when the clothesline sticks up only eight inches above the snow.)

Then there was the big journey east, followed by a cold snap and more snow. It was mid-March, and Vermont was still flaking away fresh inches in the morning. I forgot I owned sandals. I'd look at photos of Tennessee on my computer and want to cry. My new neighbors told me there was no spring like a Vermont spring, so lush and dramatic after the barren deciduous trees came back to life. But apple blossoms and maple buds seemed like something from another planet. I had been wearing my snow boots since the other side of a continent was home. It had been a very long winter.

Those following months in the cabin were physically inactive yet emotionally exhausting. I was snowbound for a stretch of weeks without a farm to keep my body, mind, and soul busy; it was starting to wear me down. Now that I no longer had a small homestead depending on me, I realized how much I missed the routines and responsibility my last place had offered. Life as a new Vermonter felt helpless and boring compared to being in Idaho. My life revolved around my desk at work and my two dogs, who had become so adapted to my world that I no longer thought about their care as any sort of effort. Feeding and walking the dogs were as routine as brushing my teeth and starting up the car. I had never longed to haul a water bucket to livestock as much as I did that February.

You can understand my need for deliberate activity. I wanted to grow again, and I mean that in every sense of the word. Adding sled dogs, chickens, gardens, bees, and rabbits to your life is about more than just rolling out of bed and shaking some food into a bowl. Your home turns into a breathing being: something that needs tending, weeding, and the occasional yeast packet. I longed to get up early and feed the rabbits and chickens before heading to the office. I even missed trudging through the waist-deep snow to refill frozen water fonts. The harmony and hardships of homesteading had completely melded into one song for me. And it was a song I couldn't get out of my head.

Maybe it was just my disgust with those still-unmelted snowdrifts, but I really missed the garden. I never set out to become a Gardener; it was on my list of skills to learn because growing my own food was important to me. A garden is a way to plant your own insurance, a way to depend on yourself for dinner even if you're cash-broke and the car's out of gas. I loved eating my own salads and stewed tomatoes, but what I really yearned for were those hot days out in the yard with a hoe, breaking sod, getting that sunburn, feeling my arms ache, knowing that a few hours of sweaty effort mixed into a heavy layer of compost and manure would produce amazing, beautiful, clean food.

I also missed the feeling of responsibility the garden gave me. Like the animals, it was another thing that needed me. Gardening is just as much of a mutual agreement of effort as raising animals is; you need to feed, and tend, and water, and weed, and do anything else to keep those plants healthy, and they'll produce for you.

I would sit in my car after work, staring out at the dark winter sky, and pull a crumpled Seed Saver's Exchange catalog from the sun visor and read through it, like garden pornography. You mean to tell me I could be growing Dragon's Tongue beans and Green Zebra tomatoes in a few months? Really? I was skeptical that the sun would ever show up again. My Idaho garden seemed like a lifetime ago. I was jonesing for some topsoil between my toes.

Come March, gardeners are all pacing like caged wild dogs. We have sacks of sprouted, mutant potatoes; packets of snap peas; and six-packs of lettuce to put in the ground. We're scratching around in our pots of houseplants to remember the feeling of working soil. During dinnertime conversation we bring up rabbit manure and blood meal, nonchalantly asking our tablemates if they think the horse owners next door would let us pick up a truckload of manure. We're awful, but we also can't help it. It's a labor so addicting, so complicated, and so dear to our hearts, we're barking for turnips by April.

To break the fever that winter, I'd grab my fiddle and work on a few songs. It was the only way to get my head out of the compost pile. But then I'd come across an old tune about shelling peas, or Jimmy Dickens would holler, "Take an old cold tater and wait!" and I'd be back to despondent thoughts of my comatose garden. That rusty fenced-in area outside just needed some swirling pea blossoms to bring her back to life. I could make out where a compost pile once was, in a crust of ice. It was sectioned off like a small wooden wheelbarrow without wheels. In a few weeks, when everything melted, I would have a better idea of what I was dealing with. My hope was that the last renters used the place to grow food, too. If they did, I might spend less time breaking sod and building raised beds and start right in shopping for heirloom sweet corn. To temper the desire to plant, I tried to think about sunburn and mosquitoes. But we all know they're just collateral damage in a life lived outside. Now in my fifth month of snow, I'd have handed the mosquitoes a damn syringe if it could have gotten me another twenty degrees and a couple of extra hours of daylight. I'm not above bargaining here.

I was overcome with this empty feeling that people get when they know what they want but can't have it just yet. Maybe you've felt that same hollowness yourself and can understand how vulnerable it makes you feel? It's no different from pining for a lover you can't hold quite yet or reading the menu of a restaurant you can't afford until your next paycheck. My recourse was simple: I put my mind someplace else.

To try to meet some like-minded locals, I posted flyers advertising beginner fiddle lessons. I'm not a professional musician by any means, but I did teach myself the basics. I also figured that because I'm far from talented and still manage to make music, I could understand the foggy beginner's mind. With my previous year of self-education fresh in my head, I thought I could help some other wannabe fiddlers out there start sawing away, too.

The flyers were a huge success. Within weeks I had met a slew of locals, all of whom were entrenched in the music scene. Some had friends in bluegrass bands, others wanted to learn to fiddle while their children took classical violin lessons at school. Some of them were even homesteading on their own small farms. A woman named Shellie from just over the border, in Hebron, New York, had forty acres with a flock of sheep, laying hens, ducks, and an old farmhouse that had been ordered from the Sears, Roebuck catalog around a century before she started farming there.

The lessons weren't professional by any means, but they did get some people playing. A few of them kept coming back through the spring and summer to sit on the porch and play. I realize now how bold it was to move into a new place and start posting flyers for music lessons. But my brashness paid off. Those sheets of copier paper were enough to get me a few familiar faces around town and some good tunes to boot. We weren't going to give anyone in Nashville a run for their money, but we were having fun, and I was feeling less like a stranger in southern Vermont.

At the office my coworkers were getting to know me. Some were slowly warming up, but in the traditional New England fashion, most cut me a wide berth. The company I work for is small, and the people who chose to live and work in a rural setting were pretty content with their lives. No one asked for a bubbly, audiophile, wannabe farmer to move into the cubicle next to them and start asking questions about local feed stores. To their credit, everyone was angelically patient. Slowly, I could see the sparks of friendship igniting with a few people.

One slow afternoon in the office while I was zoned out staring at a spreadsheet, I was awakened into consciousness by the friendly voice of someone leaning over the cubicle wall. On the opposite side of my little barrier was a young, tall, shaggy-haired blond guy who looked more like the models in our spring catalog than the men in the rod-and-tackle division upstairs. Most of them sported a two-month-old beard and had about twenty years and thirty pounds on this kid. He was another transplant, who had moved here a few years before me. He had walked down a few flights of stairs to shake my hand and introduce himself.

His name was Steve, he was around my age, and he played guitar. He told me a friend of his saw my fiddle-lesson flyers in Sandgate and wanted to know, would I like to get together and play music with them sometime? Both he and his friend, Phil, played guitar and thought that adding a violin to the mix would be interesting. I told him I'd be thrilled. The idea of making some musical friends at the office was appealing. Bridging some of my personal life and my professional life with guitar strings seemed like a grand idea.

Steve and Phil invited me into their band and their homes. Over the next few weeks, we'd spend Saturday mornings or Thursday nights in each other's living rooms, playing everything from modern covers to the occasional old-time tune. We weren't really interested in perfecting our work or renting a recording studio; we just liked to play. Phil had a young family with two children. Steve had a high-stakes job developing new products for the company. I was new to the state. All of us used the band to relax and laugh and drink a few beers, enjoying the time we set aside in our busy lives to put some acoustic instruments in our hands.

Through Steve and Phil I got to know their families and friends as well, once again amazing me at the web of connections a shared interest can deliver. Since Steve had made a lot of friends at the office, I was folded into his group and started to warm up to others as well.

Proximity breeds a wonderful level of comfort, and before long the big, tough guys who worked in the pod next to me became friends. Their names were Phil and James, both native Vermonters; they took me under their wing and became my big brothers in the Green Mountain State.

After a few practice sessions, Phil and Steve were certain we were ready to play a local open-mic night. I wasn't so sure. They had both been playing their guitars forever and were good at it. I had been playing my fifty-dollar fiddle for a few years and wasn't comfortable with their music yet. I had learned to wail on "In the Pines," not pop music. But the guys were fans of some scrappy alt country songs like "Wagon Wheel," by Old Crow Medicine Show (one of my favorites) and a few from Ryan Adams, too. They knew more about this music than I did and opened my ears to a lot of new bands. Slobberbone, Uncle Tupelo, and the Mountain Goats were the bands our nameless trio covered. We felt we had a good set of songs ready for the North Bennington music scene, so one Thursday night we packed up the guitar and fiddle cases in the back of Phil's station wagon and drove off to play at a sports bar.

I sat in the backseat on the drive to the bar and contemplated the situation. Neither of the guys seemed nervous. They had gotten their first-time performance anxiety over with the weekend before, at an open-mic night in Pawlet, Vermont. I had never performed in a band, in front of people, ever. I was as tense and nervous as a person with a fiddle case could be. I just hoped we'd get to drink first.

We walked into the bar with our instrument cases in hand. It seemed like a light and friendly place — an Irish-inspired brass pub with heavy athletic overtones: green and brown walls with tennis rackets nailed to them (you get the picture). A very loud band was already playing in a tiny corner of the already tiny bar, right next to the bathrooms. My first-ever gig would happen about seven feet from a urinal. We set our cases along the wall and ordered drinks.

At the bar were some friends from work who'd heard we were playing. It was nice to see familiar faces, but also nerve-racking. It's one thing to mess up a gig among drunken strangers but another thing altogether to mess it up among sober witnesses polishing off their plates of pasta — especially sober witnesses you'd see at the office in a few days.

I was really starting to second-guess myself. I was nowhere near as talented as Steve or Phil. And to top it off, my instrument didn't have any way to be connected to the speakers, so chances were slim that my little fiddle would even be heard amid the ruckus.

Our name was called off the open-mic roster, and we took our places on stage. There is a moment of pure excitement and anticipation in getting set up to play music in front of a crowd — the plugging in of cords and amps, the arranging of band members and microphones. If you're careful and pay attention, you can pause to take in the audience's expectation; they'll always give you the benefit of the doubt. I took my place on the far left side and tried to place the sound holes of my fiddle right below the high-volume microphone hovering over it. I was standing there in front of the packed crowd, staring out at friends, when Steve started the first chords of "Wagon Wheel." My shaking hands began to saw out the opening fiddle notes. We were off.

The actual song was a blur, but I knew if I focused on what I had memorized, at least I wouldn't mess up the other guys. I stared at my strings and tried to make sure I hit all my marks. The boys were singing the choruses and I was singing along with the verses. When my solo came up, I played the simple drone notes as best I could, loud and proud. Someone in the audience put his hands in the air to clap. I had never been more grateful to an inebriated individual in my entire life.

When I dragged the last note of the song across my fiddle and the crowd gave a pleasant clap, I exhaled for what felt like the first time all night. Someone yelled, "Turn the fiddle mic up!" and we did. After a few more songs, we took our place back at the bar and celebrated with a round of Guinness. I was still shaking as I sipped the tan foam. Shaking but pleased. I had done it. Barely anyone could hear my little violin over the wail of the people and guitars, but I did play it, in public, in front of people. It probably didn't change anyone's life or even come up in conversation later that night, but *I* wasn't going to forget it. I had taught myself to play, made friends, practiced with a small band, and legitimately performed in front of a live audience. It was its own form of cultivation and felt just as amazing as eating my first slice of homemade bread or growing my first tomato in the backyard.

SHEEP 101

I RETURNED FROM MY LUNCH BREAK at work one day to find a folded copy of the *NewsGuide* at my desk. The *NewsGuide* is our local weekly paper, the kind of publication that announces garage sales, classifieds for used motorcycles, and minutes of the Catamount Rotary Club meetings. This particular issue had a University of Vermont Extension class circled in red Sharpie ink. I picked it up and started to read the announcement some agricultural fairy godmother had left for me: "Beginner Sheep Raising Class. Held in various locations throughout the state. UVM livestock specialist Chet Parson will instruct this daylong workshop in basic shepherding." The cost was minimal, and the location was only forty minutes north of my cabin in a farm education center called Smokey House.

As I stood there reading, a friendly voice rose up from behind me.

"Did you see that class?" It was Trish, a fellow designer in the office. We'd gotten to know each other through a mutual love of fiber and knitting. I was still a beginner compared to Trish, who not only knit but also spun her own yarn. Through our chats at the office, she'd learned about my farm dreams and knew I wanted my own flock. She had left the paper on my desk, thinking it would be fitting for me, a place to start scratching my itch and see what really went into raising a few sheep in New England.

"Are you kidding me?" I gushed. "I'll absolutely take that class. . . ." This was perfect. Just a few months in Vermont and the state was handing out flyers for my dream lifestyle. I signed up online, mailed my check, and was told to show up early on the Saturday morning of the workshop.

A few weeks later, on a sunny, cold morning in late March, I found myself at the Smokey House Farm Center, sitting in a metal folding chair in a classroom lined with maps and farming posters. About twenty people, of every age, shape, and size imaginable, were milling about, calmly paging through their handouts. Next to me were a couple in their mid-sixties and to my other side were a couple in their late twenties. I seemed to be one of the few single people in the room — possibly the *only* single person. Did I read the fine print wrong? It didn't say anywhere on the website that this was some sort of couples retreat ("Better Relationships Through Second-Cut Hay Purchases"). Apparently, sheep bring people together.

I was feeling slightly self-conscious about the fact that I had come without a date. I'm not ashamed to admit I scanned the crowd for a solo thirty-something guy in rumpled clothes with a border collie at his feet and a steaming cup of coffee in his hand. (Better luck next time.) My wishful thinking was quickly interrupted by a crisp, friendly voice from the front of the room, announcing that we were about to get started. I looked up at what could only be Chet, our instructor.

Chet was a graying, bearded fellow in a ball cap. He stood among us like a high school coach prepping his team before the big game. He *exuded* experience . . . exactly the kind of guy you'd want around three months into raising your first flock, when something unimaginable went wrong. He spent the next three hours pontificating about wannabe farmers, expounding on the legacy of New England shepherds, and giving us what I can only describe as a field guide to sheep breeds. He covered the popular breeds for our region — the hardy meat stock, as well as handsome wool breeds popular with the hand-spinning market. Chet also went through the sheep I was most interested in: the dual-purpose breeds — sheep you can either wear or eat, depending on your preferences.

When it came to raising a small flock, my intentions were fairly humble. All I wanted was a small group of healthy animals, maybe five or ten, I could raise for both the closet and the table. The dual-purpose breeds like Romney, Cormo, Rambouillet, and California Red would offer wool of spinning quality and still pack on enough pounds to fetch decent market prices. I wanted to raise breeding ewes that would drop lambs in the spring for market and drop fleeces, too. I wanted enough animals on my someday farm to have a supplemental income and keep me stocked with lamb chops, but not so many that a single woman with the help of a good dog or two couldn't maintain them. This class was making my notion of keeping sheep seem like a real possibility. I squirmed in my seat.

All around me, people who were just as excited and curious were asking questions and talking about their farms. They asked about what kind of fences to use, where to set up water tanks, and if it's true you should buy goldfish for your stock tanks to eat algae. I'd never even thought of that! I drew a goldfish by my notes, while a recently retired couple explained how little they wanted to mow their lawns and hoped a pair of sheep could do it for them. We all had our reasons.

Then it hit me: Most of these people already owned land. I was squatting on someone else's backyard and trying to will it into production. I tried not to let that get to me or dampen my hopes. So what if I didn't have ten acres and a green tractor? I had a pasture (kinda). And I had a landlord who already allowed dogs and chickens. Maybe she'd be okay with three sheep and a small pen, some portable fences, and a shed. Raising sheep on rented land wasn't *that* crazy.

As the class continued, Chet told us that at one time Vermont was Sheep Central. Long before my great-great-grandmother hopped a boat to America from Presov, Czechoslovakia, 80 percent of my current state was pasture and 20 percent was woodlands. (Now those numbers are reversed.) Apparently, a fellow named William Jarvis signed an agreement with Spain to allow Merino sheep to be raised in the state of Vermont. Napoleon was wreaking havoc in Europe, and the Spanish didn't want the breeding lines of their native sheep in danger of being depleted. Thanks to that mess, Jarvis was able to start a flock at his famed Weathersfield Farm (from the original English word *wethersfield,* as in a castrated ram's field).

Sitting in a historic building, hearing a shepherd talk about his livelihood's history, surrounded by people as intoxicated by the idea of lanolin on their palms as I was — it was pure heaven. I decided that if I ever bought a ram from Vermont, I would name him Jarvis. My thoughts were interrupted when someone handed me a taxidermied rumen — a sheep stomach hollowed out so that we could see what we'd be filling up with grass in a short while. It smelled like old soup.

When breaks came between lectures, I sidled up to people and introduced myself. I figured we were already in a sheep class together: clearly, we had an icebreaker. I talked with an ex–broadcast journalist who raised Sebastopol geese and dreamed of a flock of rare Soay sheep. He and his wife were here to learn what starting that flock would entail. Another woman sitting to my right had always dreamed of having a fiber flock. After years of spinning other people's wool, she wanted her own.

Another young farming couple I met already ran a meat CSA. They were new to sheep but had seventy lambs being delivered *tomorrow.* For that couple today was not an afternoon of speculating about hobby farming; it was a crash course. I had never been so jealous of anything quite so terrifying. Tomorrow this happy couple would be rounding up dozens of lambs. I would have done anything to be in their place. Now I knew why they were asking so many questions about pasture rotation. It was what they'd be doing Tuesday afternoon while I was in an e-commerce meeting. Some days you sigh deeper than others.

There are a lot of ways to keep sheep, but the main options seem to fall into three categories. Some people just raise spring lambs for the table; this keeps things simple and cheap, since summer grass is free and winter hay is not. Alternatively, you could have a full-season flock that you overwinter, shear, and live with all year but do not breed. And then there are people like me, people who hope to experience the entire spectrum of ovine thrills with a "fully fleeced flock," as it's called. I would be buying lambs, rotating pasture, breeding with stud rams, buying hay, and becoming a midwife.

All of that sounded perfectly logical until Chet started sharing the complications of lambing. Now we were seeing the darker side of shepherding, the real stuff. Apparently lambing goes as planned "99 percent of the time," Chet assured us with a grin, but there were a few "situations" we needed to know about. He showed us some of the many ways a lamb could be presented to the world incorrectly. The screen behind him showed lambs being pulled from the womb by their back legs and shepherds in elbow-length gloves going inside a ewe to organize hoof placement, turn bodies, and (I am not making this up) tie ropes and pulleys to akimbo limbs to pull out a stuck lamb. Then there was the additional complication of twins (which is common) and even triplets.

I watched, fascinated, but increasingly nervous. I had just assumed those little guys plopped out on warm grass one day and hung out with Mom until I saw fit to separate them. Not the case. Some lambs die, some get rejected or forgotten by their mothers, some ewes die, and sometimes all of the above happens, and if you don't have a heat-lamped lambing crate ready for orphans and a vet on call, you could be in a world of trouble. I kept taking notes, nodding, and asking questions. I was not going to waver. The lambing complications did not deter me as much as they made me shake my head and hope I'd never be in a predicament that requires twine and lubricant.

I was a little shaken, but still excited, when we broke for lunch. After we ate we'd be heading down to the barn to get our hands on some livestock. I walked out into the March sunlight and stretched long and lanky as I could.

Even though the temperature hung in the mid-forties, most of us picnicked out on the grass. Bundled in sweaters and scarves, we talked about our future flocks. None of us knew each other personally, but the solidarity of shepherds is strong. There were no real strangers here. We all wanted sheep in our lives, so we were all comfortable with each other. Sitting Indian style on the grass, munching a peanut butter and jelly sandwich on bread I had baked the day before, I felt very pleased with the world.

After lunch we stood outside the red, weathered barn waiting for Chet. Everyone was playing his or her own version of cool, sitting on the outside steps of another outbuilding or leaning against a car, but we were all dying to get inside that barn and get some wool between our fingers. Some of the other participants were able to mask their excitement, but I was having a hard time hiding mine. Even the most level-headed among us turned their heads at every stray *baa* or whenever the hired, coveralled farm help opened the sliding doors.

I can't speak for any of the other students, but walking into that barn was like crossing a different kind of threshold. My knees started to buckle, just a little. This wasn't my barn, and these weren't my sheep, but this was the first time I was ever such a hands-on participant in the life I wanted to take part in. I didn't have a partner, or land, or a barn, but I was there. And standing in that barn was 100 percent closer to my own flock than I had been the day before. And I was miles closer than I had been sitting in a Borders café in college, paging through a copy of *Hobby Farms* magazine and daydreaming about livestock. When I couldn't believe it was really happening, that I was standing in a sheep barn in rural Vermont, I could reach down and feel a lamb's ear in my hands. It was happening. Consider me pinched.

Inside the barn, dusty shafts of sunlight revealed fifty ewes and their spring lambs. Multicolored babies skittered and played tag at our feet while their patient parents chewed cud and occasionally *baa*ed at us if we got in their way. Some of the more nervous ewes skittered to the wall, which was lined with a long trough of hay.

I noticed that the trough was made with slats in different sizes, to enable ewes, lambs, and their guard llama to eat at different places and avoid piling on top of each other. It was simple but ingenious. I found myself taking more notes in the barn than I had in the classroom. I jotted down how hay was presented and where medical supplies were stashed. I asked the farm help where they dumped mulch straw. Or did they prefer the deep-bedding method?

During all the barn demonstrations, there was a little ram lamb that wouldn't leave us alone. He was a little black guy with a white blaze on his head. Every time I stopped petting him, he started chewing on my jeans, so I crouched by him and let him lean against my side while I listened to Chet. He was like my own private fan club, a nice little boost when you're covered in sheep manure from the knees down.

When Chet asked if anyone would help him catch some ewes to demonstrate handling and foot care, I hopped the fence before anyone else could even get a hand in the air. Together we cornered three or four ewes, and when one tried to race out past me, I crouched and caught the hundred pounds of panic as if I had been doing this my whole life. "Nice catch," was all Chet said. I beamed.

We spent the rest of the day handling the animals, watching Chet's demonstration of the proper way to flip a ewe on her rump, tag ears, crop tails, and inject medicine. I kept my pen hand busy, but most of this was the kind of experience you need to practice firsthand to gain any real competency. That couple getting seventy lambs tomorrow would be able to put all this know-how to use right quick. I realized with some sadness that it might be years — maybe even *decades* — before I could tag lambs of my own.

After all, there was a reality to my limitations, both financially and geographically. I was a low-level corporate employee living on my own, so my income was modest. Although I was thrilled to be using Photoshop at a day job in a town where I could be attacked by a puma, it did limit my possibilities. I couldn't buy my own farm like some of these other classmates already had. I didn't even rent a farm anymore. In Idaho I'd had land I could grow into, dozens of acres, even if it wasn't mine. Here in Vermont, I lived in a small cabin, and of the acres I had, only half an acre wasn't covered with trees. So, even if by some miracle my landlord would allow me a few sheep or a goat, space was still an issue.

When the barn work was done and we were all back in the classroom, my mood lightened considerably. Before saying thanks and a final "Good luck," Chet handed each of us a small yellow ledger that looked a lot like a checkbook. It was a lambing record book. It had the University of Vermont logo and a Suffolk lamb outline on the cover. It was a simple record-keeping device and probably cost no more than forty-five cents to print and bind.

It was nothing special, yet I held this little yellow book like a golden ticket. I sneaked a sly glance at the other attendees to see if anyone else was as excited by it as I was. They didn't seem to be affected. But how could they not be? We were holding a gem, the talisman of a culture that 99.9 percent of Americans would never know. This was a tool possessed only by people who walked in their fields, bought hay by the truckload, and knew more breeds of sheep than of dogs.

As far as I was concerned, it was a passport to the future. I'd taken the class, shared a meal with shepherds, and held a scrambling ewe in my arms today. Now I held a little notebook I would someday write in with bloodstained fingers in a dark April barn. It was my proof that I was one of the few, the hay-stained, the happy.

THE ARRIVAL OF RUFUS WAINWRIGHT AND BENJAMIN FRANKLIN

AS MUD SEASON DRIED UP, the state bloomed into the green Xanadu my neighbors and coworkers had promised. The contrast of stick season with the now flowering apple trees surrounded by a carpet of green seemed to happen overnight, as if one day I was driving home from the office and the landscape suddenly woke up. This green all around me was what I missed most when I was living in the West. The Rockies were beautiful and dramatic, but so stark compared to all of this. My northeastern soul preferred these gentle rolling hills, covered bridges, and lush pastures. I think I might be part Hobbit.

Even as the outdoors started to look warmer, though, there was no mistaking that it was still early spring. Nights dipped into the teens, making it feel more like early January at times. But that hint of green fueled my desire to enlarge my flock. There was plenty of room to house more egg-producing poultry, and I didn't think a pair of rabbit hutches could hurt either. I made plans to attend the Annual Poultry Swap at the local fairgrounds the first weekend in May. The fact that it coincided with the annual Rutland rabbit show was kismet of epic proportions. Maybe I'd get some laying-age birds and a pair of pedigreed French Angora rabbits and start my own small rabbitry. I had raised them in Idaho; now I could breed them in Vermont.

Every year the first Sunday in May caters to an unusual crowd at the Schaghticoke Fairgrounds. Just a skip over the state line into New York and you'll find yourself at what equates to a livestock tailgate party. It's called the Annual Poultry Swap out of stubbornness for tradition. At one time it was just the local chicken club swapping pullets, but it had long since evolved into the kind of marketplace rarely seen in America today. By eight in the morning the grounds have been transformed from an empty dirt parking lot into a small festival of fur and feathers. True to its name, it is a swap meet. Small farmers and animal breeders come with everything from turkey eggs to two-week-old goat kids, with a mind to sell, barter, or borrow.

The makeshift roads are lined with the open sides of vans and pickup trucks. Wooden hand-painted signs and card-board posters leaning against trailer tires list options and prices. Want a pair of Silkie bantams? Looking for a ram lamb to add to your flock? Need a basketful of goslings or guinea hens? How about Angora rabbits or a litter of beagle pups? They're all here. (I've yet to see someone up the ante and bring in a pony, but it's only a matter of time.) I'm telling you, this swap meet is a ripsnorting hayride for the livestock set, and I make sure I'm there every single year.

I heard about the event from a local newspaper reporter who was interviewing me about my first book. Since there was a chapter in it dedicated to chickens, we got around to talking about establishing a new flock in Vermont, and she asked if I knew about the event. I didn't even know swap meets like this still existed, much less that there was one thirty minutes from my cabin! Apparently, word of mouth was the only way news of this event got around. I couldn't find anything online anywhere, and none of my local Vermont papers listed it. (If they have since that year, well, then, to all rural journalists of New England, I apologize.) I heeded the two words of advice the reporter gave me: Go early.

I showed up before seven, and mine was one of the few cars in the parking lot, but behind me a steady stream of people was percolating into the fairgrounds. I walked into the agriculture building and spotted an older man with perfectly conformed golden-and-blue Seabright bantams. If fairies were to reenact *The Empire Strikes Back,* bantams are the kind of tiny beasts they would ride — tiny, fancy tauntauns. The next man over had Dutch rabbits. A woman past him had boxes of chirping babies: day-old chicks and turkey poults. Another man was setting up goslings near a quail pen. And these were only the first exhibitors. Outside, the parked cars formed rows and alleys, truck beds turning into Nubian goat shelters and cage stores. A clever gardener set up a stand of vegetable six-packs. He had the market cornered and could barely keep up with the shouting customers.

I walked past his booth, turned a corner, and spotted a small horse trailer backing up. A crowd gathered, excited for what was inside. Children jumped up and down and middle-aged women looked on with wild grins. I pricked up my ears and heard a baker's dozen of tiny *baa*s and nickering. A small gate was raised around the back of the truck, some cedar chips were spilled, and out of the back hatch flew five kicking goat kids and right behind them a pile of baby lambs. There is no animal on this planet more adorable than a two-week-old goat kid.

I watched the babies romp and play and the people haggle over prices for the stock inside. The woman was a local farmer and was asking fifty dollars a lamb or kid. I thought that was fair. If it were a pet, fifty dollars is a heck of a lot less than the cost of a poodle. And if it were destined to be dinner? Well, you just bought what would end up being a hundred and fifty pounds of freezer meat for the price of three large pizzas and a two-liter soda at Pizza Hut.

I was there to experience the festivities, but I was also on a specific mission: I needed a rooster for my flock. I had already picked up some laying hens on a road trip near Lake Champlain, but as yet there were no men in the picture. Hens don't need a rooster to lay eggs, but they're happier in the company of their male counterparts. A rooster's job is to protect and rally his flock. He's part guard, part lover, and part caterer. The roosters I've known have always kept a watchful eye over their ladies, performed their role as a mate (a lot), then spent the rest of their free time finding delicious things to eat and yelling for everyone to come check them out. My hens were laying just fine and didn't ask for my dating services, but for my own peace of mind I wanted a security guard on duty. I was hoping to score a nice young cockerel for under twenty bucks.

Usually, finding a spare rooster isn't a tough thing to do. People in the country are always giving them away. The rural Craigslist Farm and Garden posts were teeming with mistakenly gendered "laying hens" looking for new homes. I'd seen the ads and could have answered them, but shopping at the swap meant I could peruse at my own pace and have my pick of the gentlemen. So whenever I walked by a cage or truck with a rooster, I'd let my gut reaction and my wallet have a conversation. Most of the roosters were older and scrappy (calling them "pot stock" would be putting it kindly). Others were beautiful but were either fancy bantams (I don't think a Silkie bantam rooster can even reach a Jersey Giant's *wings*, if you get my point) or giant Cochins sold in pairs. I didn't need a rooster as big as a toddler: I had only one hen he could service without putting her in traction. I kept on looking.

Then I noticed a teenager pulling crates out of the back of a van. His business was clearly the beautiful wooden cages he was selling to people who needed to take home new animals, but he also had a few birds for sale. Among them was a red-and-gold rooster with a long green tail. I recognized the breed right away; my mentor in Idaho raised these birds. It was an Ameraucana. His grayish green feet were the dead giveaway, but other standout features like cheek feather fluffs and those wild party colors made me certain of the identification. I asked the boy how old the rooster was, and he said all the birds he had that day had recently turned a year.

My instincts were rooting for him. I asked, “How much?” The boy seemed shocked that I wanted to pay for a rooster but stuttered out, “Fifteen dollars.” I handed him a twenty and he handed me a five-spot and seven pounds of squawking feathers. I held the bird close to my chest. He calmed down and rested against me. I carried him through the crowd to my station wagon, where I had a small cage ready in the back seat. As I walked through the throngs of onlookers, some people commented that I’d found the best-looking rooster at the swap. I gratefully agreed.

When I got to the car, I loaded him gently into the cage and suddenly realized he needed a name. I watched him strut and move around the small pen with style. He might have been terrified, but he was playing it cool. He bobbed his head as though he was keeping time with a song in his head. I couldn’t get over the music in him, and his long green tail feathers seemed like lavish coattails. He cocked his head and looked up at me with his tiny yellow eyes. Then he firmed his footing, ruffled his feathers, and busted out a long, smooth croon. This rooster wasn’t just crowing; he was singing. Suddenly, the perfect moniker popped into my head, and I rested my hands on my hips, satisfied with the decision. I named him Rufus Wainwright, after the musician

I'd come to admire, whose own style and voice seemed to fit this wayward bird.

I woke up early the next day and drove north to Rutland. The Long Trail Rabbit Club was holding an American Rabbit Breeders Association (ARBA)-sanctioned rabbit show, and at this particular show, I was going to pick up my breeding stock. The animals I hoped to attain were French Angoras — a beautiful breed of fiber rabbits that grow wool so dense and soft, it's been called the warmest on earth by people in the biz. The breeder was a woman from Massachusetts named Nancy Platts, who specialized in this breed. She said to meet her at the rabbit show at 9:00 a.m., so we could go over pedigrees and care and feeding before I took my pair home.

When I pulled into the Vermont State Fairgrounds, I could not believe the cars. Rows and rows of rabbit people were parked there, with license plates from all over New England that read **MINIREX** and **SHORABTS**. I guess rabbits were a bigger deal than I'd realized.

Inside the ag building were hundreds of cages and rabbits (surprisingly, it smelled fine), and along the walls were stewards taking notes and judges in white coats examining every inch of their subjects. They'd feel their heads and feet and talk aloud about their bone structure and bite. I watched a few judges work with a breed called the Flemish Giant, a huge rabbit with a satin short coat, then walked across the building to see the Jersey Woolies — small, long-haired bunnies — being judged. It amazed me how different rabbits could look from one another. Like a lineup of purebred dogs, each had its own purpose and history.

I went back to Nancy and sat with her by a judges' station, while she went through the two rabbits' pedigrees with me. The buck I was buying was a big guy with light brown wool. I named him Benjamin Franklin because he looked like a Ben, and who doesn't love that sassy character? The doe was a fawny cream color called Lynx and had giant brown eyes. I named her Bean Blossom after my dream banjo of the same name.

The plan for the two fluff balls was to raise them to breeding age and mate them. If all went as planned, I would have my own purebred rabbit kits in the backyard hutch. Since I would be getting into the business of breeding rabbits, I joined the ARBA and, with the help of more experienced locals, would be bringing some adorable little Angoras into the world by midsummer. Along with the income from selling bunnies, I'd be selling their fiber and spinning some of my own.

Later that spring I made a split-second decision in a feed store that would end up causing trouble come November. I had gone to Whitman's Feed in North Bennington to pick up my order of poultry: six light Brahmas, six Ameraucanas, and two Toulouse goslings. The chickens were all pullets — laying hens for my egg supply. I bought the geese as a pair, simply because I liked them. I had always liked geese, and finally, I was living a life where "What the hell, throw in two

French geese with my order; I'll just bring a bigger box" was a normal sentence.

I walked into the back room of the feed store and happened upon four giant, waist-high wooden and wire cages, filled with chicks basking in the soft glow of warm heat lamps. Day-old chicks are ridiculously adorable; you hold them in your palms and feel their heartbeats in their talons. They cock their heads and pivot tiny black eyes around the room. I held my fourteen new beings one at a time, saying hello, listening to their soprano voices, which sounded the same even though they were different species. *Om* may be the sound of the universe, but *peep* is the cry of all future poultry.

As I headed to the register with my boxful of livestock, I spotted a sign that read **EXTRA BROAD BREASTED WHITE TURKEYS! $5!!!** A little enlightenment seeped into my brain. I made a decision right then and there in the cacophony of chirps and the orange glow of the feed-store heat lamps that I would be taking home a turkey. I'd give it the best life I possibly could, and then, after a few months under my care, he'd become part of my family's Thanksgiving dinner.

The plain fact was that, come November, my family was going to eat a turkey. I'm the lone vegetarian in a family of fervent carnivores (my grandfather was a butcher). They would not be swayed by a flaccid serving of Tofurky, nor would they abstain from eating meat. Since they would be eating a bird regardless of my preferences, they had two choices. They could go to the supermarket and buy a turkey that had lived in some factory farm or they could opt for something that had lived naturally, outside in the sunshine.

In a few months my turkey grew from a poult that was smaller than a side of mashed potatoes into a cute, quirky, creepily smart teenage bird scampering in the grass with laying hens, sled dogs, geese, a duck, and a new litter of crip-plingly adorable Angora rabbits. He ran freely among my menagerie with a languid jog — a tiny velociraptor on lithium, his white baby feathers almost gone from his leathery pink neck. Sometimes, when he wanted to impress the hens he grew up with, he puffed up his pathetic teenage plumage, lifted his tail, and actually looked something like the turkeys you see on television. The hens weren't impressed and walked away, bored. His vulnerability was endearing.

INTO THE GARDEN

STARTING A VEGETABLE GARDEN is a marriage of high-stakes poker, ardent independence, and possibly incurable masochism. The gamble of weather and new seed packets, the free food that (at times) makes you certain you could feed a Civil War battalion, and that sick need to shove your hands in dirt and stay out weeding after sunset are all well known to those of us who grow our own french fries. I am one of these people. Hand me some poker chips and call a therapist, folks: I am a gardener, and it cannot be helped.

You start out one year with the idea that you can grow a little food. You don't have anything extreme in mind: a modest bed of tomatoes and maybe some herbs for your pasta. So you plant a few varieties of heirloom reds, and a few months later — biting into one of these girls you pulled right off the vine — you realize that the taste is something you can't believe was the work of your own hands. The basil makes the back steps smell like a Tuscan countryside. When you need some onions, you just pull them out of the ground. Suddenly, it hits you: Not only does this gardening business taste good, it's also making dinner a lot easier to plan and your grocery bills fewer and farther between. It works, and you're hooked. You decide next season to make some modest improvements. You'll add another raised bed because you heard stuffed zucchini was to die for and you want to grow your own pumpkins for Halloween. You close your eyes and remember the hayrides to the pumpkin patches of your childhood and how great it was to pick your own jack-o'-lantern from a farmer's field. Okay, two beds. See where this is going?

Little changes start to happen, things normal people barely notice. Only your closest friends may start to see the signs of addiction. They can see the unapologetic potting-soil stains under your fingernails at a dinner party, or the shopping bag from the local nursery in the corner of your kitchen when they're visiting for a cup of coffee. You start getting seed catalogs in the mail, and your borrowing history at the local library shows a suspicious number of agriculture books. You start setting aside a beat pair of jeans for sod breaking.

Then the real addiction kicks in. A few years down the road, your basement has been converted into a nursery and the aboveground pool has been bulldozed for more raised beds and a grapevine. You've hired a carpenter to build a series of containers to grow more strawberries, and there are messages on your machine from orchards, seed savers, and the people from your vegetable-gardening meet-up group. Suddenly, you know famous gardeners by name and can even drop a few in conversations. Then, late one July day, you're in the grocery store checkout line and notice the woman behind you looking closely at your basket. Meat, eggs, milk, flour, bananas, pectin, pickling spices, vinegar, sugar, and three boxes of pint jars. "You're a gardener, too, huh?" she asks. You nod, smiling. And you see the same collage of ingredients and supplies in her cart, all produce lacking, because there's a good chance she has that section of the grocery store in her backyard, too. You get to chatting and find out she has the tomato variety you read about in last month's issue of *Mother Earth News* and you have those white pumpkins she's been coveting. Friendship is struck, numbers are exchanged, and the nice people with plastic bags full of trucked-in vegetables are none the wiser.

And then the big change comes. One day over breakfast, you look across the table at your spouse. Or you *would* look across and see him eating his steel-cut oats, but you can't because your squash harvest is in full swing and there is a small mountain of zucchini in the way. Regardless, you kind of cough to clear your throat before asking, ever so casually, "Hey, honey, what do you think about getting a place in the country?"

I was one of these people now. I'd never intended to be a gardener, and certainly not one with fifteen raised beds, a small corn crop, and a swirling pumpkin patch, but it happened. And while I had yet to ask a significant other to uproot for greener pastures, I could easily see myself doing so. I couldn't imagine a life without a garden. After the summers I'd spent growing vegetables in Idaho, and knowing how good the food tasted right from my own backyard, I was already starting to make plans for my Vermont garden in early March.

And did I ever have plans for this place. The cabin's garden was ready to be worked, and I was champing at the bit. Finally, after months of ice and snow, seed catalogs, and online browsing, I had come to some decisions about the layout, vegetables, and planting schedule. Lacking a tiller, I chose to stick with the same semi-raised-bed method I had in Idaho. I say "semi-raised" because I dig the soil about eight inches deep, removing stones and roots along the way, then pile on about nine cubic feet of compost for each five-by-eight-foot bed and mix it in. I edge the beds with whatever's around: scrap wood, fallen tree limbs, old license plates, or broken tool handles. These beds have served me well. So I planned on filling the entire garden area with as many beds as I could. I would place stones and sunflowers between the raised beds; I would make even smaller plots and plan for herbs and strawberries, too. It was going to be like walking through a Whole Foods produce section — no, better, a farmers' market! — in my own backyard. I was going to start with the basics, though. I had lettuce, broccoli, and peas to plant that first planting day in spring, and it would start with a ravenous hoe.

I raised up my favorite hoe (well, my only hoe) and slammed it into the dirt. With my iPod blaring and the morning sun cresting the hollow, my body got warm and happy with the purposeful work. I had broken sod before and now knew what to expect. Sweat started to bubble on my forehead. Within the hour my arms were screaming and my blue bandanna was soaked around my ears. Gardening has taught me to care little about discomfort in the present. Sore arms and stinging eyes are so temporary. With every thrash and grunt, I could already feel the hot shower, smell the mint soap, and feel the clean clothes I would wear later. And as I stood over those rows of freshly planted starts, I could sip my hard cider and smile.

Growing your own food isn't easy, but it's wonderful to earn. Every bite of future salad is enriched with the education of days like this.

This garden gospel is easy to spread to the seed-saved, but what about when a nongardening friend from New York City comes to visit for a weekend?

My friend Nisaa sent me an e-mail saying she'd been thinking of me and was eager to hop a train to Brattleboro to spend the weekend. She wanted to visit the cabin and get her hands dirty. She'd been reading my blog and thought my homesteading adventures were intriguing, a feral alternative to her city life. She wanted to come and help me put in the gardens, settle in the new laying hens I'd bought off Craigslist, and see my new neighborhood.

I invited her with open arms but was a little concerned she'd be miserable. Not so much with me, but with my life. I was the same friend but in a different world now. After all, this was a lot different from Manhattan; I didn't even have cable. Would she really want to come perform manual labor for food she wouldn't get to eat? It had been years since we had spent time together, and when we did, it was in her city. Why would she want to come out to the sticks and play in the dirt? Was she sure? Apparently, she was. We made our plans, and she bought her Amtrak ticket. Away we go.

Nisaa and I had gone to the same college in Pennsylvania and had become fast friends, but we couldn't be more different. Nisaa, quite frankly, has style. She is into fashion, celebrity, music (we have that in common), and probably knows more about pop culture than the editors at *Rolling Stone* do. She is also black, and raised Muslim, and grew up in Philadelphia. I am from Palmerton, Pennsylvania, a place so white that the 2000 census indicated that minorities made up less than one percent of the population. As for religious and ethnic diversity? I think the Greek Orthodox church at the edge of town was as exotic as our town got. We didn't even have a synagogue.

But despite our different backgrounds, Nisaa and I clicked. On car rides and in classes, we had each other laughing till our stomachs hurt. I adored her. I just didn't think she'd be into getting chicken shit on her shoes (which, by the way, would be far nicer than mine).

In a few weekends she was by my side, hoeing in the garden. It was hotter than usual in Vermont that May. We both cursed at the bugs and checked our hands for blisters between our fingers. I could tell she hadn't expected the work to be so exhausting. Bugs were swarming. Our heads got foggy. Like all people in uncomfortable and close quarters, we got short with each other at times. We silently fumed at the rocks and roots. But to her credit, she kept at it. She worked hard and dove right into the dark soil, ripping out sod with her bare hands. I could not believe this was the same person who had called me from Strand Books just days before. She had gone a little feral herself.

I told her we should take a break. I had a water station set up on the porch with a five-gallon container of springwater and some lemons in Mason jars. But she shook her head. Her wonderful stubbornness kept her hoeing and ripping up clods of soil. I watched in silent awe while I worked beside her. Eventually, we did take a break and walk down to the stream to soak our feet in the cold water. As we sat there in the shade of the woods, we rinsed off our foreheads and arms and talked a little. She and I were both ready to quit, but the sod we were breaking still needed more work. And after the soil had been turned, I wanted to mix in some compost and get the lettuce starts in the ground.

We both let out a long sigh as our feet splashed. We knew this job wouldn't be over until it was over and that the garden wasn't getting planted while we sat in the shade. We groaned happily and got back to our feet, a little shaky. We were both out of shape, and the garden was making that point all too clear. It was just a five-by-five-foot salad patch, but it had two strong adult women ready to pass out. We sighed and walked back up the hill to the sun and the dirt. Back to work.

It took another hour or so, but we got the thing planted. We made little rows and lined the edges of our beds with scrap wood, and eventually it looked like a garden. After showering and having a few beers (and, thus, in a slightly better mood), we strolled back out to the garden in the dusk. Nisaa stared at what we'd done. I watched her watch the earth. In the middle of this shoddy, fenced-in, scrappy pile of turf was one beautiful garden bed. All around it was ugly sod, long-forgotten rusting fences, and a broken gate, but in the center was this promise of clean food. The baby lettuce shoots in the black soil shone like jewels. For an amateur job, it looked good. Nisaa crossed her arms and smiled: "I get it now."

After Nisaa's visit, every weekend (and some weeknights) became a date with my hoe, compost, seed packs, and trowel. I'd tie a bandanna around my head, braid my hair into pigtails, and hike out into the garden in my busted coveralls with a smile much akin to something roller-coaster fans may display outside the locking-bar gates. I always carried music to work with. Usually, it was my iPod, loaded with playlists appropriate for sod breaking. (A lot of Radiohead went into that first garden. If I ever meet Thom Yorke, I'm going to tell him all about how great my *OK Computer*–inspired arugula was on pizza.) By late April my arms stopped hurting and picking up the water buckets seemed easier. In a few weeks I had planted fifteen raised beds. I could stand at one end of the garden, look over my little plant minions, and feel genuinely wealthy.

I am comforted, even if it's just a little, by my garden and flock of hens. Knowing that there is a free source of protein and vegetables right outside my door brings me a little security at a time when the prices of gas and grain and the world's shortages of food are all I hear about on the radio (that and the wars). While I think it would be tough to survive for a *long* time on what my little homestead produces, I know I can make at least half of my meals during the summer with food from the backyard. It's a lot of work, but I also save a lot of money. A six-pack of broccoli plants costs me $2.79. Depending on the season, a head of organic broccoli at my local grocery store costs $3.49. Growing my own seedlings from a $2.75 pack of seeds is even cheaper. And I can always keep planting more if the skillet calls.

It's not simply about collecting eggs and harvesting vegetables, either. There is comfort in the other skills I pick up along the way. When strawberries are in season at a nearby-farm, I can make a year's supply of jam in one afternoon with just a few dollars and some Mason jars. With a few pounds of flour and some yeast, I can bake all the bread I can eat. A good tomato crop will give me all the pasta sauce I can stand through the winter. If I'm lucky, I'll have a few jars of golden honey and some homebrewed wine as well. Knowing how to produce, preserve, and create some of your food feels pretty good when the average price of a barrel of oil is heading northward with no signs of stopping. If there was ever a time to start learning to garden, it's now.

I'm not a conspiracy theorist, and I don't expect the recession to drive us into a depression. I don't think all of us in America need to turn our lawns into victory gardens. (I do, however, strongly believe we'd all be happier if we did. It's harder to be angry at the news when you're biting into your own roasted and buttered sweet corn.)

Homesteading, and gardening in particular, is the best way I know to be rich without spending a dime. You might be swiping a credit card at a giant grocery store and eating expensive food till your gullet is overflowing, but that's just plastic and gluttony. If the power goes out and the store's cash registers won't run, you're not eating a grain of rice. In that same blackout, though, you can walk out to your backyard with a flashlight, eat a fresh organic salad for free, and watch the stars.

True wealth is not about money; it's about independence. Gardening gives you back that basic freedom. And that all might sound a little cliché, even to the greenest of Prius-driving Sierra Club members, but it's true. Want a chance to make a million dollars? Sign up for the stock market and throw your hat in the ring. Want to *taste* a million bucks? Plant some snap peas; when they're ripe and covered with dew, bite into a pod and start chewing. Now *that's* rich.

MEET THE LOCALS

GETTING COMFORTABLE IN A NEW PLACE gives you a little confidence. Knowing names and faces at the feed store, in the parking lot at the office, and around the hollow made me feel slightly assimilated to my new Yankee Life. I was starting to pick up on some of the local sayings, like "Sorry about it," which is often used in place of "Screw you" but makes the person cutting in front of you at the grocery store to grab the last head of broccoli seem mildly empathetic to your plight. This was pointed out to me by my coworker Andrea, who had experienced the same little vernacular trick after moving from Connecticut. Nobody is sorry about anything. It's simply something people around here say to acknowledge that they just ripped off another living being. I'm pretty sure deer hunters say "Sorry about it" after they pull the trigger, too.

Yet not all my realizations about this new place were negative; in fact, few were. And those that were only seemed to add to the color of the area's social topography. Most of the new experiences were quite wonderful, actually. Mud Season lived up to its name but also brought Maple Syrup–Tapping Season. It seemed like the time of year that scared away 97 percent of the tourists was the best time to be a New Englander. I'd drive to work and the roadside sugar maples were connected with lines of plastic tubing, collecting sap into giant metal tanks. It looked like the whole forest was hooked up to a delicious IV. I was able to buy some syrup from Merck Forest, a local operation that sold several grades. The woman behind the counter explained to me that locals eat grade B and the tourists eat grade A. "We like the taste of maple in our sugar water," was her pithy explanation. I bought two quarts.

The staff at Wayside, the local general store, started conversations with me about gardening and introduced me to neighbors with chickens. I started to settle into the place and into my life as a small-scale-homesteading, day-job-working Vermonter. I learned there were a lot of us, too. Other folks at work had chickens and bees; a few had horses or sheep. Everyone (and I mean *everyone*) seemed to have a garden.

But even in the most rural of places, there are extreme differences among the citizenry. While some fault lines exist in politics or religion, the biggest divider between neighbors seems to be class. I began to notice it more and more. I'd be standing at Wayside, two gruff men in matching red flannel jackets with bushy beards ahead of me in line to pay for their Sunday coffee and paper. We'd talk about the things that bind us as country neighbors: the weather, our livestock, the price of gas, and what we read on the front page of the paper.

Then the lot of us would walk out to our cars, and suddenly I'd realize the difference between the two men I'd assumed were members of the brotherhood of agriculture. One walked over to his new Toyota Tundra and the other tried to open the frozen door on his rusted twenty-year-old F150. It hit me how different these country neighbors' lives truly are. I homed in on the details, noticing tiny giveaways I hadn't put together in the coffee line. The one getting inside the $43,000 truck was wearing top-of-the-line silicon Muck boots; the other wore the rubber ones that cost $19.99 at Tractor Supply. The Tundra had a purebred golden retriever in the front seat and the Ford hosted a small bird-dog mutt curled up on duct-taped upholstery. One man had retired here to become a gentleman farmer, raise some horses, split wood, and fill his days with fly-fishing and deer hunting. The other was barely getting his dairy out of the red. They both lived in the same town, shared conversation over the same cup of coffee, but their lives were utterly different.

The truth is, money silently rips people apart. It's never spoken of in public, but it's strongly understood. You notice it the most when bad things happen. The power goes out, and you lose your running water and heat, yet your neighbor with the 134-acre property starts up a generator so loud it wakes your dogs. You understand immediately which side of the coin you're on. While you're busy wrapping pipes and splitting firewood for warmth, he's watching a WWII special on the History Channel with a hot toddy in hand.

I'd recently heard through the vine of Sandgate gossip (which flows from Wayside in copious amounts) about the neighbor's new barn. It was the talk of the morning, this giant structure that was going up in a few weeks. Everyone had a guess what it would be for and when it would be completed. The news about the construction shocked me, mostly because the owner's current barn was already seven times the size of my house and provided plenty of space for the dozen or so horses he owned. A second barn? *Really?* Couldn't he add stables to the current structure? And why did a man who traveled so much and was so rarely around need another enterprise? To start raising dairy cattle? Sheep? A pig barn for raising his own salamis and charcuterie? None of these turned out to be the reason for the construction crews assembling the gigantic structure that fall. He was building a barn to house antique cars, with a recording studio upstairs.

It was a slap in the face — unintentional, and nonc of my business, but a slap nonetheless. It was like building a recreational baseball field in your backyard when your neighbor learned to pitch by throwing rocks at trees. The idea that an ark was being assembled at great cost to match the rural landscape and to house a man's hobbies made my stomach burn. The land that family owned was for leisure and for show, but if my dirty-fingernailed fists ever got hold of such a gift, what an empire I could carve! Sometimes I would walk by the property (just off the main road I walked with the dogs every day) and shake my head at the lost possibilities of the plantation. I could grow enough food to feed the whole town. I'd have a pair of workhorses and turn those fields into meat, milk, and wool production. I'd have acres of vegetables. I'd start my own vineyard. I'd grow wheat and corn. And I would love every second of it. I'd think about this and sigh and walk by with my dogs. It was all the worse that the owner was a chef. He could live an entire life around food if he wanted to; instead, he played with it. Anyone who doesn't think decadence thrives in America is fooling himself.

Not far past this empire lay another example of personal/agricultural heartbreak I encountered on my regular dog walk: the blue cottage. About a mile down the dusty road from my mailbox sat a small blue house nestled by the side of the road. Across the street were a small barn, a pond, willow trees, and old pastures. Beside the house itself were a small horse shed, an acre of pasture, and a well-maintained sugar shack previous owners had used to make maple syrup. The house was well kept, humble, and nothing compared to the hundreds of acres and several large structures of the horse farm and car barn down the road. Yet in a way this place was much harder for me to accept. The owners, whom I'd met on such dog walks, were from an affluent suburb outside Philadelphia. This was their second home — a place they occasionally visited when their calendars were clear. At least the guy blaring Grateful Dead covers in his barn studio was around to feed his horses on the weekend and pick some tomatoes. But this young couple had made a dream homestead into a rarely inhabited vacation destination. As someone who could have turned that plot of land into so many things in one year — a sheep farm, a rabbitry, fields of organic gardens, a waterfowl pond, an orchard, a honey factory, and God knows what else — seeing it domesticated into a gentrified weekend getaway actually nauseated me when I walked past. I had to tell myself "It's just not my time" and speed-walk past the always empty house with a lawn mowed too short for the deer to bother with.

These are the kind of people a lot of agrarian Vermonters deal with. The rich guy playing farmer, the second-home owners ignoring their property. And I'm sure I would have fit into the taxonomy of annoying Vermonters myself. My green-living, composting, sustainability-driven small farming probably annoyed the hell out of the conventional dairy farmers trying to scrape by. Seeing a girl with a decent-paying job throwing her money into a backyard farm so she doesn't have to buy the conventional food they're struggling to produce must make them shake their heads. I saw how I was to them. But intentions and desires are complicated things. I had every right to grow my own ice cream if I wanted to. Car barns are not illegal, and neither are vacation homes. We're all just trying to live the lives we aspire to.

A wiser version of myself understood that stereotypes are pointless. And around here they often weren't so clearly defined. At the feed store I may have run into a yuppie ski-home owner buying organic dog food for his Weimaraner as easily as I could have struck up a conversation with the cheese-making goat farmer posting signs about Nubian kids for sale. The dog-food buyer may have had his deer license and been born in Rutland, and the goat farmer may have been an ex–New York novelist, and still I'd find myself more similar to the guy who knows how to milk a teat than to the guy who spent every Christmas break at Killington. The intentional lives we grow into is where we most often find common ground.

In Sandgate the residents find common ground in its small-town traditions. There's the Christmas party with Santa at Town Hall. There's the dead-Christmas-tree bonfire and children's sledding night every Martin Luther King Jr. weekend. There are pig roasts and parties, but regardless of the season or the guest list, they all center on the age-old practice of sharing home-cooked food.

Potlucks are king here in the sticks. Since the community covers the catering, it requires only a friendly host and some sort of beverage to get the party started. Somehow, no matter how many you go to, potlucks never get old. How could you grow tired of something that requires such little effort for the amount of fun you end up having? You make one dish, grab a six-pack of beer, and show up to a spread of mismatched tables covered with food. Potlucks are unapologetic in their gluttony, a chance for each of us to share our special talents. (My neighbor Phil makes the best cheesy potatoes you'll ever eat.) Tables overflow with so much food and finger desserts that you get self-conscious standing in line. We may not have a single stoplight in our town, but we sure know how to feed a mob.

Every summer in Sandgate there is a community ox roast, and this was my first one. Around dusk, I left the cabin for the ox roast with two very important things in hand: a pie and my fiddle. Generally, people who show up with fiddles and pie are welcome in almost every enjoyable place in America. This is a truth to live by, friends. If these two items aren't welcome where you spend your free time, you're hanging out with the wrong crowd.

So when I crested the steep, rocky driveway of the farmhouse, I knew instantly that the night would be pro-pie and pro-fiddle. Sprawled out before me were old Colonial buildings and a big white barn. Picnic tables with fresh flowers dotted the lawn. Sandgatians smiled and nodded as they sipped iced tea from Mason jars. The twilight sky was lit by table lamps on wooden pillars or set high in barn windows. (Extension cords were the workhorses of this fine evening, that much was true.)

All around me were hundreds of adults, kids, and the occasional dog running around off leash. In the center of the commotion were three musicians in red plaid shirts playing a fiddle, a guitar, and an upright bass. They were sawing out a version of "Blackberry Blossom," a beloved old-time fiddle tune. My heart swelled.

These were my people now: a feral group of New Englanders who square-dance in tie-dye or tap their maples in stoic red plaid. They're farmers, loggers, small businessmen, bookkeepers, and florists — pretty much any odd job that lets them be the boss of their own lives. But most of all, tonight they were a happy, wild-eyed people who wanted to be outside with their neighbors instead of inside with their televisions. For that, I wanted to kiss them.

This was not a group of people who drive their garbage bags to the curb and don't know how the people next door pay their mortgage. This was a *community,* and this newcomer was going to get to spend a night getting to know it a little better. It was a bona fide first date. My mom always asked me if I was "seeing anybody" because she hated that I was still single. Well, call me a hussy, but that night I was on a date with the whole 247-year-old town. I stood there in my old hat, holding a cast-iron skillet of apple pie, a fiddle over my shoulder, and walked into the fray smiling. I told myself, "Men will come in time, but for tonight let there be food and music!"

Food and music there were! The smell of a steer on a spit filled the air, and a huge potluck spread filled rows and rows of tables. Twenty yards of cloth-covered folding tables adorned with every form of Tupperware and Pyrex made since 1964. There was a giant canteen of iced tea and an outdoor freezer sporting our local hero's product, Wilcox Dairy ice cream (which was what all southern Vermonters ate; since Ben & Jerry's ice cream is from *northern* Vermont, it is not local enough).

Of course, there was also a full cast of characters, live and in person, like the maverick genius from Washington, D.C., who wired up the UN's first phone service. People said he drove his neighbors in the suburbs crazy with his antics and backyard projects. He belonged in Vermont, one older lady said, as her flock of old-lady friends nodded in silent approval. She said this as matter-of-factly as if he had broken a leg and needed a cast.

I spent most of the night listening to stories about the people who lived here. My favorite was about an original Norman Rockwell painting that someone found in his deceased parents' house, jammed behind a false wall. And there was the story about the two women who built the Sandgate covered bridge by themselves! I listened wide-eyed and enamored.

How did I end up in this amazing town? What Fates helped me find my cabin in a random want ad from twenty-eight hundred miles away? By pure chance I landed here. I felt blessed.

As the sun went down and my stomach was full of good food and maple ice cream, I pulled out my fiddle. My neighbor's beau, Sam, and I played music while other people digested. Simple guitar and fiddle tunes in lonesome chords. We stopped when the paid band started up again. Slowly, people made their way to the dance floor, which was lit by a Tiffany-style lamp hoisted up by a ladder from a tractor. People twirled around while the string band's bassist called out square-dancing moves. The local kids knew all the words to "Red River Gal." There is hope for America yet, I tell ya.

We stayed for a few more hours, mostly to talk, sip wine, and laugh over stories about work and family. I left pretty late, but folks were still dancing when I pulled away in the station wagon. I was happy. It went well, as first dates go. I was developing a serious crush on this place.

THE SOCIETY OF LAMB AND WOOL

AS THE SUMMER TURNED into a stretch of farm chores and fly-fishing excursions, someone at the feed store told me about the sheepdog trials over at Merck Forest. Apparently, every summer there was a big open event held for competitors around the entire Northeast. It was quite the to-do in the local world of competitive herding, and since it was less than fifteen minutes from my front door, I had to check it out.

I did a little research online and found the club hosting the trial. It boasted a fully interactive website for herding enthusiasts, complete with classes, seminars, beginner trials, photo galleries, and a membership program with a newsletter and lending library. I had no land of my own, no sheep, and no sheepdog, and my entire rented cabin seemed to be somewhere around the size of a holding pen for ewes at such competitions. And yet something about all this resonated with me; I had to be involved with it. I had wanted sheep for years, had dreamed of understanding and experiencing that relationship between a working border collie and a handler. For so long I had nurtured all these big plans for "someday." But when fate dangles a carrot in front of you, you just have to bite.

I clicked on the information for the Merck Forest Open Trial and e-mailed Steve Whetmore, a sheep farmer in upstate Vermont who was listed as the main contact for the trials. I didn't know if that meant he was running the show or that he was the sucker who had to field e-mails from crazy people like me, but it went something like this:

Hello, Steve,

My name is Jenna, and I'm really interested in getting involved with your club. I don't have sheep, or a sheepdog, but I will soon. Can I come watch your event, volunteer, or do anything of use? How does one get started in all this? —j

He replied shortly thereafter, telling me to just show up and introduce myself, that he'd be happy to talk to me more about the collies and the trial. I was beginning to picture myself standing on a windy hill with a black dog and a crook. I'd be in a waxed-cotton coat and slouch hat, and if people didn't notice the white ear buds, they might think it was 1846. Sign me up.

The trials were held on a miserable Sunday morning. Or rather, the *weather* was miserable. It was midsummer, but it might as well have been October to those of us outside on that angry, gray day. The rain was falling sideways. The high grass all around me was slick and cold, whipping around in the harsh wind. But I'd happily left my warm bed at dawn to be a part of the eager crowd on that rainy hill. All of us had arrived to watch fast paws and panicked hooves.

Despite the weather, I was having a grand time. I pulled down the brim of my felt hat to keep the mist off my glasses. In hopes of staying a bit drier, I sat hunched with my back to the gale, secretly wishing I had one of those snappy folding chairs everyone else seemed to have brought along. Lacking such luxury, I sat on my backpack instead. Honestly, though, I wasn't too concerned about the weather. I was transfixed on the athlete performing in the pasture below.

Roughly a hundred yards from my vantage point, a border collie named Cato was herding Romney sheep. His black-and-white silhouette darted and dashed. I'd never seen anything like it before. He seemed to sidestep between raindrops as he expertly followed his handler's whistled commands. I barely blinked, not wanting to miss an instant.

Cato herded his small trial flock down the hill toward his shepherd. I'm not sure what modern shepherds are supposed to look like, but this fit, middle-aged woman was wearing a technical mountaineering jacket and jeans. If it weren't for her large crook and the triangular silver whistle around her neck, you'd have thought she was posing for a Patagonia ad. Cato tore into the green around the post and somehow got the sheep to circle around it and head back the way they'd come, driving them back up the hill. I knew it was a border collie's instinct to gather and herd sheep, but how did that dog get trained to lead them away? I leaned forward to take in more, as fascinated as a child at her first drive-in movie. It was eight-thirty on a rainy weekend morning, and I was grinning like an idiot watching these sheep. Normal people do not do this.

Fortunately for me, "normal" in this kind of a place was hard to define. Sheepdog trials aren't like suburban soccer games — the subculture is scattered, underground, and made up of people from every corner of society. Like the woman at the post, the members look surprisingly ordinary. These were not elders in Donegal tweed; these were baby-boomers in baseball caps. The average crook-wielding participant with a border collie at her heel could have passed for a dental hygienist or a landscape architect.

For some reason, this pleased me. Seeing these people leaning on their crooks next to a dented Dodge pickup and not a vintage Land Rover made the whole notion of becoming one of them more reasonable. They seemed to be walking, breathing examples that this can be done — anyone can decide to become a shepherd. A sheep person may be ahead of you in line at an ATM or she may be teaching your kids advanced literature in their freshman year of college. They're Red Sox fans and Yankee fans, plumbers and poets, farmers and financial consultants. Some came into the work of tending sheep by choice, and others were taken there by their dogs. More than one unassuming border collie owner has ended up in sheepherding classes with his overactive pet and before he knew it had a small flock of dog-broke ewes in his suburban backyard. Others have always lived among the fleece, knowing the feel of lanolin on their palms since before they could walk.

I wasn't any of these people yet. I wasn't even the clueless spectator with a pet-shop border collie in my arms. I was just a girl with a hunch.

After two days of watching, eavesdropping, and asking complete strangers stupid questions, I knew that Cato had just accomplished a beautiful out run, executed a fine lift, run down a straight-lined fetch, turned the post, and completed the drive, and was in the middle of a decent cross. He had penned and was about to shed. His handler, who someone told me was a former rock climber, yelled commands into the wind. When the wind proved too much, she blew into the whistle on the lanyard around her neck. Cato balanced the sheep like a pro. Not too far from where I perched, a man in his mid-sixties stood in an oversized blue sweatshirt, and on the end of a frayed lead, his border collie sat beside him. They were next. I was so envious of him I sunk a quarter inch into the ground. I had come on a lark, didn't know a soul, and barely understood enough of the sport to keep up with the sideline conversations wafting around me. But I was happy. I didn't even mind having a soaked butt. I was falling in love.

The idea that I could make a living raising sheep for wool and lambs, work outside with these amazing dogs, and perhaps write about it from time to time made my palms sweat and my voice pitch higher. These people were living my dream. Regardless of who they were or what they did with their lives, they all had managed to figure out how to become shepherds in the modern world. They all spent their days with ewes and rams and these highly skilled dogs. I wanted to be one of them so much it hurt, but like the dogs waiting their turn to pump up the hill to their flocks, I needed to be patient.

Even though every muscle in my body quivered to get a border collie and a couple of sheep as fast as I could, I needed to be realistic. Finding a way to own a farm and transition to a full-time farm career seemed almost impossible. The hundreds of thousands of dollars in a mortgage, the start-up capital, the high credit score, even just the electric fencing seemed far out of my reach.

I also needed to remember that I already had two roommates at home who might not adapt to the shepherding lifestyle. I'd arrived at Merck dogless. For someone who owns two wonderful working dogs and was going to a working-dog event, this felt wrong. But I didn't have the heart to bring Jazz and Annie. Besides the fact that the temperature the day before had been in the 90s (scorching hot for two huskies with heavy undercoats), I couldn't bear sitting there with two dogs who had to be held back on leads while they watched countless other dogs scamper around leashless inches away from the animals they desperately wanted to devour.

I could visualize Jazz shaking his head at those sheepdogs. Like an old Baptist preacher watching wayward youths rob a liquor store, he would surely be despondent seeing his own kind having fallen so far from the faith. Siberian huskies keep the Gospel of Wolf alive in every fiber of their being. I could just imagine myself on the trial field with my Siberians, yelling "Away to me, Jazz! Come by, Annie!" and the crowd would see the fastest sprint clocked in sheepdog-trial history as my sled dogs ran toward the flock. They would be in awe at the grace and beauty of my dogs as they loped with the bliss and agility of Russian ballerinas rolling on Ecstasy. And then they'd scream in horror as Jazz and Annie ran down and began eating the ovine contestants.

Sigh.

So the dogs were at home. I felt as though I was cheating on them, and all I was doing was window shopping. I wanted to be a shepherd and I was living with wolves. It was a complicated situation.

There's a reason I've chosen a wolf with antlers to represent Cold Antler Farm. It's because he's the symbol of how I feel about my passions — the mix of opposites, the chaos of contradiction, the need for balance when you love things that are so different from each other. And the possibility of seeing a wolf with antlers in the wild felt as unbelievable as the possibility of my becoming a shepherdess with a trial dog.

While all this was churning in my head, I bought a shepherd's whistle at the merchandise stand. A small souvenir to some, but to this girl it was the crucial key to a life I knew I wanted but had no idea how to attain. At the moment it seemed like all the start I needed. Maybe someday a wolf with antlers really would trot by the farm. I'd just have to wait and see.

Coming home to the cabin after a whole weekend of sheepdogs was jarring. I'd been given a taste of a life I felt an intense connection with, but I was still limited by my current circumstances. Leaving the trial grounds felt like breaking up with someone I knew I could love.

My mind was made up. I was going to be a shepherd. I knew this with the same certainty a kid playing in the summer sunshine has that school will start again in September — it was what systematically happens next.

Somehow I was going to become a shepherd, and it was going to happen right here. I didn't want to put it away in my mind and wait for some perfect moment or some perfect home. I don't see the point in waiting for things to happen to you or hoping the perfect circumstances arrive. I wouldn't jump into the world of livestock without the proper research or intentions, but I wasn't going to wait for the stars to align either. My backyard might be small and scrappy, but it could grow wool.

The first step: find sheep, then ask for permission to raise them. The second step: build them a proper shed and pen. I have this theory that if you are constantly thinking about something, it finds a way to wiggle into your life. Kind of the same way that if you start thinking about an old college friend long enough, someone else brings him up in conversation or you get an e-mail from him, as though he heard the dog whistle of the universe.

After a month of driving around New England to visit herding workshops, see beginner trials, and watch private lessons, I was starting to obsess a little. My mind was woolly, and I spent my spare time learning whatever I could. I joined the Northeast Border Collie Association (NEBCA), ordered

DVDs from its library, and watched them on the futon in the cabin after long days in the garden. My life was becoming about three things: the office, the farm (such as it was), and becoming a shepherd. Eventually, sheep found their way to me.

One day during our fiddling hour, my friend Shellie brought up the idea of a barter — I would get a few of her sheep in exchange for giving her lessons. My heart pounded. I told her I *absolutely* wanted them but I needed to get permission from my landlady first and set up a shed and pen for them. That same night I asked for permission and got the good word: yes.

Within a few weeks of catching ovine fever, I was getting the green light on a trio of my own. I was going to become a shepherd.

The backyard farm I was creating was starting to look less like a backyard and more like a farm. The chicken hutch now housed adult chickens, my pair of gray geese, a few ducks, and a fat turkey. I had installed a hive of bees and built a compost pile behind the rabbit hutches. All this occupied an area roughly the size of a basketball court surrounded by a dead-end dirt road. The garden had exploded into more than a dozen raised beds of varying sizes. The neighbors had gotten used to, and even said they enjoyed, having the chickens wander down to their driveway or the geese float in their stream. My little farm was a free-range poultry wonderland.

I was excited to be graduating to hoofstock. In my mind, the farmer I wanted to be lived and worked with larger animals than chickens and rabbits. She had a goat on a halter and a whole pile of sheep munching on grass around her in the pasture. She had a working horse that plowed and pulled a cart and maybe a few cows on the hill or a pig in her barn. I was grateful for the wool bunnies and egg layers I already had, but a girl's got to dream big. And this part of the dream seemed to be coming true.

THE HOOVES HAVE LANDED

I HAVE NO IDEA WHAT IT'S LIKE to wake up the morning of an actual barn raising, but the morning a carload of friends were coming to the cabin to build my sheep shed felt special. It must have had the same air of purpose and community as a real barn raising. Early on a Saturday morning in late summer, a car pulled up to the farm with volunteers, tools, and extension cords. We had no plans, only a rough idea and some spare wood and roofing.

I was glad to see them — relieved, actually. See, I'm not handy. All the tools I now possess I've picked up from the hardware store because friends who came to help at the farm asked for them so many times. Hand me a leash or a livestock halter, and you'll see some competence. I can tell good hay from poor hay. I can set up a flock of chickens in a new coop; just don't ask me to *build* the coop. My skills are sloppy. I live by Joel Salatin's motto, "The pigs don't care if the feeder isn't straight." Functionality always wins over form. So my "construction jobs" are cob jobs and rarely pretty. Usually they don't last as long as they would if they were built by someone who understood water runoff and knew how to use a level. That said, I'm also not a complete idiot or a debutante when it comes to carpentry. I'm not scared of the hammer. I do try. But my nonmathematical brain makes everything crooked and wonky. So when something as important as a four-season shelter for my first trio of sheep was in the works, I wanted something solid. I asked for help.

At the office I approached my friends Phil and James, who had become my closest confidants in the state. Knowing my level of skill with power tools (and lack of them), they offered to come help build the shed if I got all the pieces in one place. I could handle the staining, fences, and gate installation once the actual shelter was constructed.

I was excited! All summer I had spent my weekends driving around New England, going to sheepdog clinics, workshops, and trials. I had been spending most of my nonwork time either in my garden or among shepherds, who had become agrarian superheroes to me. I was still shocked that being a shepherd was a viable career choice in the twenty-first century. These people found a way to make herding a part of their modern lives, worked with and among animals every day, and were bringing real products — food and clothing — into the world in a wholesome way.

It was a side effect of all this time around meat farmers that I started doubting my vegetarianism, too. Sheepdog trials and herding clinics are places to share recipes and roast lambs on spits. It was a part of their culture and soon to be a part of mine. I made a mental note that if I ever went back to the world of meat eating, it would be lamb that would take me there. Sheep were responsible for the wild ride I was currently on and would be responsible for many more: border collies, buying land, knitting from my own wool. It seemed only proper that they'd take me to a place at the dinner table as well.

And so, on that sunny morning, the three of us and my neighbor Casey (who supplied most of the tools and all the lumber, which were floorboards he'd rescued from an old Arlington sawmill), set up our sawhorses, plugged in our extension cords, and started measuring out boards and planks. Hearing the ruckus, my neighbor Roy sauntered over in his signature sweatpants and T-shirt. He saw all the tools and lumber and asked me what I was planning on building, because he didn't want something unseemly ruining his view. I actually laughed out loud at this, thinking he was joking. He looked serious as a health inspector who's just found rat droppings, so I changed my tone. I mumbled something about it being tasteful, and that I would be getting three sheep, and he walked away looking fairly upset.

It never occurred to me that people in an agricultural/woodland community like Sandgate would get hung up on the aesthetics of an outbuilding in a neighbor's backyard. Actually, I was shocked at the idea that someone would be more concerned about the physical appearance of the building than its purpose. I would be much more concerned about "why" the kid next door was putting up fencing and a small barn than about how it looks. This assumption was, of course, utterly naive. When he was out of sight, Casey (who was holding a circular saw and talking loudly over it) told me that Roy was from New York City, that what he saw from his back deck was of grand importance, and that I should expect this from any flatlanders coming into the country. I scoffed. I guess country living was okay with them as long as it had curb appeal.

With all the distractions and spectators behind us, we got to work. I was given the task of nailing the plywood wall to its bracing posts, and did my level best to nail it straight. It was still a tad crooked. (My day job is safe.) James shook his head while he marked the planks with a pencil for cutting. Phil was busy digging holes to bury the cinder blocks at varied depths, so that the building could be level on the uneven ground. Casey set up the sawhorse and ran power from the chicken coop's humble night-light. Soon conversation dwindled to work talk and simple requests. Hand me that. Cut here. Are there any more screws? How long is that piece of plywood? We fell into the rhythm of the work, and I did my best not to get in the way. I tried to stay mindful that these people were giving up their time and resources to help make this happen, and I was truly grateful, but I lost my serenity the third time I hammered my thumb into the plywood.

Realizing that nailing-only jobs were few and far between — most of the work now lay in setting planks on the frames, using a circular saw, and measuring things properly — I went inside to make everyone lunch. I had a warm loaf of bread from earlier that morning, sun-ripened tomatoes from the garden, and some local cheese. Making do with what I had in the kitchen, I offered everyone a cast-iron-skillet panini and some lemonade.

While we all leaned against Casey's yard tractor and surveyed what we were about to complete, I took in the scene. With four sweaty people, an old sawmill, and some cordless drills, we had built a shed large enough to hold three sheep comfortably. It had three solid walls with a ventilated, slanted roof and was raised up on cinder blocks (off the wet ground). Its front side was only half open, meaning the beasts would have near-total protection when the winds howled or heavy snow fell. It still needed a fence, a gate, and some staining, but those would come soon enough. I sipped my lemonade and thanked everyone. It had taken only three hours to come this far. I beamed.

Thirty-six hours after the first power tools were plugged in, the pen was complete. We built the majority of the simple structure that morning. The next day I stained it, pounded fence posts, put up fencing, installed the gate, and built a hay feeder. The final touch was a real metal roof, a gift from James's father, who had bought too much for a home project and had a piece exactly the right size for the shed. James came over with Phil to put the final screws in the roofing, and it was done. Right there in my backyard were a tiny barn, fence, gate, and water and grain bins. No sheep graced it just yet, but they would soon. Everything seemed too pristine to be real. The gentle green moss. The slightly turning leaves of the sugar maple that hovered over the new roof. The straight and perfect fences. The crystal-clean water in the black-plastic trough. It was poetry.

By the time the job was completed, my eyes were welling up. I'm not normally one to cry at wooden structures, but this was an ark. There was no way I could've done this without these people. Like so many things in the self-sufficient life, the more deeply you get involved, the less self-sufficient you become. I needed people more than ever before. Learning to build a sheep house took help. And not just help with the construction, either; it took a willing landlord, gifts from neighbors, a delivery from the fencing company, trips to buy T-posts and screws — all of it took a small army to give three woollies a nightcap and a hotel. Four brave souls spent the weekend at my homestead making it happen. With their generosity, sweat, and gifts of wood and metal, my sheep would have a home. My training-wheel farm was causing community to happen.

The night before I was to pick up my sheep, I took the two red halters I'd bought and hung them over the fence. I stepped up onto the swinging metal gate — a *real farm gate,* like the one my friend Diana had on her ranch in Idaho — and leaned my body over the top rail. The shed still smelled like the near-black stain I had coated it with. The grass inside the pen was soft, inviting. The water trough was filled, and a mineral lick was waiting. All this shepherd needed to do now was get those three fine animals into this pen. This was something of a logistical complication, as I didn't have a truck or even a trailer. But I had a plastic tarp and a Subaru. Farmers had started with far less.

I folded down the backseats of my Forester and spread the tarp over them. The cargo area was roughly the same size as a truck bed but painfully shorter. I'd read about people transporting sheep in cars before, but usually they were lambs in dog crates. I, on the other hand, was on my way to put four hundred and fifty pounds of ovine goodness in the back hatch of my station wagon.

I had traded three adult sheep for fiddle lessons: ten lessons each. They had been part of a hobby flock, and with winter coming the owners, Shellie and Allen, wanted to save on hay costs. They were thrilled to make the barter and I was even more thrilled to accept it. I'd first met my new flock the weekend before at a picnic. We were already on a first-name basis. The three sheep were Maude, Sal, and Marvin. Maude was the sole purebred, a Border Leicester ewe with an ear tag and papers. The others were wethers (castrated males) and were a hardy mixed breed — Border Leicester crossed with Romney. Both breeds are renowned for their wool, coveted by hand spinners. As someone eager to get her hands on a spinning wheel, I thought this trade was looking better and better.

When the car was loaded with its tarp, halters, and a small stepladder (which I'd rigged as a gate between the cockpit and the hatch), I pulled down my hat over my ears and hopped into the car. I was driving over into Hebron, New York, a small farm town a few back roads from Sandgate. I was so excited, I can barely remember anything about the trip there. This was a culmination of years of hope and research and a summer of rain-soaked sheepdog trials. I was miles and minutes from becoming a shepherd.

I do remember pulling into Shellie and Allen's farm. It was a gorgeous sunny day, and their four-year-old daughter, Lucinda, all curly hair and tan skin, was standing barefoot on a rocky outcrop above their driveway. She laughed and waved to me as I pulled up to the house. Shellie came out in her muck boots and pointed to where I should park. As I backed my hatch toward the sheep-pen gate, I felt my hands shaking a little. This was it, Jenna. This was it.

We used an old dog leash and the halters I had brought to pull the sheep into the back of the car. I thought it would be more difficult than it was. We discovered that sheep like moving forward, and if the two front feet were set up in the back of the station wagon, they would accept being lifted in the rest of the way with little fuss. All of the sheep complained a little, but within moments we had the trio *baa*ing in the backseat.

We all agreed that the short ride (only twenty minutes, if that) would calm them. Originally, the idea was for me to pick them up and take them home one at a time, but this seemed to work fine. Sheep are simple creatures and generally roll with it, as long as they're surrounded by familiar creatures. They stood, stomped, and then pooped copious amounts before finally lying down on the plastic tarp.

We closed the hatch door and shook hands, and Lucinda peered in the window to say good-bye to Marvin, her favorite of the group. I thought this would be heartbreaking for her, but she was all smiles and waves. She knew she'd see them at her next fiddle lesson. I thanked the family for the billionth time and turned the station wagon out of their driveway toward home.

As I drove, three big, woolly sheep faces loomed in the rearview mirror, their red halters contrasting with their white fleece. I caught an eye and winked. The sheep stared back at me, uninterested in my flirtation. I had no idea who was who but I was beginning to learn some visual cues. I knew Maude had the bald head and no spots. (She was also the only one with an ear tag.) Marvin had a cut left ear and a spot on his front leg, almost like the brown knee pad a gardener would wear. Sal had a big mop of curly hair on his head and lacked tags and spots.

Every now and then, they rattled into a bleat fest as I turned a corner or rolled down a window. "Relax, kids," I'd semi-whisper to them as Sam Beam played on the stereo. "You're going to a great place. And although I'm going to steal your outfits in the spring, you'll eat well. And I built you a hell of an apartment." They still just stared. I guess we'd work on our communication skills later.

Actually, I wasn't as worried about our repartee as I was about their dinner. My coworker Nadine — who had a flock of Texas Dalls in Hebron — had given me a few small bales of hay as a congratulations gift for getting my own stock. I had a list of potential suppliers of second-cut hay pinned to the fridge, and Nadine had offered to sell me more as well, so it wasn't like these guys were going to starve anytime soon. Also, the farm was loaded with high late-summer grass. I had a few weeks of free salad bar already lined up for the new tenants. But winter was certainly coming, and a Vermont winter is no joke. I still had to find a local hay farmer, and soon.

Sometimes you get lucky.

As we crested the hill up Hebron Road, heading back to Sandgate, we passed a farmer and some of his employees loading bales of beautiful green hay onto the back of a pickup truck. The hay looked absolutely amazing; it was so green and lush, I was ready to pour some vinaigrette on it and dig in myself. Could I be so ridiculously lucky as to run into a hay dealer on the drive home with my first livestock?

I slowed down, and the station wagon hit a bump in the pavement. Everyone *baa*ed. That caused the hay makers to look over at what they'd previously thought was a normal car. They didn't do a double take (this is Washington County, after all), but they did shake their heads and laugh. To them I was just another flatlander who had moved to the country and bought some sheep as lawn ornaments for my second home. Or maybe they thought I was crazy.

I rolled down the window and leaned my head out to yell over the sounds of the motors. Despite the ruckus, the sheep in the backseat remained calm. They had either accepted their fate or were plain bored. All three were lying down with their elbows tucked under them, like Zen monks. They watched as I called up to the older gentleman running the show on top of a hay truck. "Hi, there!" I yelled. "Do you guys have any hay for sale?" Without missing a beat, the older farmer (he must have been eighty-five) retorted with a belly laugh, "Do you have any sheep?" Then we were both laughing at the absurdity and within minutes were shaking hands and making introductions.

His name was Nelson Greene, and he'd been farming on this same spot in Washington County his whole life. He had inherited the farm business from his father and had kept it going ever since. Independent dairy farmers are a dying breed, and Nelson was one of the few American dairymen still among the living.

I let my eyes dart around behind him as he was talking. His farm was huge, and the view was breathtaking. His world of milk and hay sat on the edge of a rolling hill, perfectly sited to take in the setting sun. I was now only half-listening to his story. I snapped back into the conversation when Marvin (I think) belted out some slow jazz behind me.

Eventually, I told him something about myself, but my story wasn't half as interesting. I told him I'd moved here from Idaho in the winter and so was new in town. I also explained that these were my first sheep (he grinned and shook his head) and that I lived over in Sandgate. I asked if he'd be around for a while working, because I'd like to come back and buy some hay tonight, if possible. He told me to drive back after I unloaded the animals and he'd fill up my car as best he could. I drove away from Nelson's farm, convinced that the world had been folded on dotted lines for me this day. Life can surprise you with its tiny generosities, especially when you're carpooling with ruminants.

When I got back to the cabin, I parked the station wagon as close to the sheep pen as I could manage. Then, one by one, I took each sheep by the halter and placed it inside to feast on grain and get acquainted with the new digs. Sal balked a little but with a firm pull and a kind word he trotted through the gate. Marvin was next and was even easier to unload.

I was foolish to assume Maude would be as easy. No one named Maude would be easy. I slowly opened the hatch and reached for her red nylon halter, cooing softly to her, as if I did with the boys. She looked back at me with wild eyes, like this was the drop-off from the last train into a slaughterhouse. When I had the lead in hand and the door open, I gave her a gentle tug. She leaped past me out of the car, almost ripping my arm from the socket as the halter jerked. Then, to our collective despair, the slip halter slid off her muzzle and over her neck, turning the gentle lead into a noose. She was confused, bleating, and pulling to get free. I held on, worried she'd bolt for the woods and be gone forever. I needed to get her into that pen.

In a panic, breathing heavily and gasping, she fell to the ground. I was at her side immediately, talking to her softly and removing the halter from her neck. She was helpless on the ground, as all sheep become once they're on their backs. (This is the trick to shearing them; turn them on their rumps and they're about as feisty as rag dolls.) When she was breathing normally, I grabbed her by the wool on her shoulders and above her rump and acted on gut instinct. I led her into the pen, and she ran to be with her friends. I swear she turned around and leered at me. I apologized, telling her she didn't have to freak out like that. It would be months before she let me touch her again. As it turns out, sheep remember everything.

When the sheep were in their pen, and looked like they weren't going to form a SWAT team and vault the fence, I headed back to Hebron for the hay. The day was nearly over, and the sun was getting tired. I knew that when I finally got home, it would be dark. As I headed back through the winding dirt roads to Nelson's farm, I turned on some music and got lost in the reverence of the day. I was already nostalgic for the present. The music was soft and lovely, and I was overcome with the emotion a met goal inspires. I had sheep. I was, in some sense, a shepherd. Tears started to fill the corners of my eyes, and I began to sing along with the music. It was on that first ride to pick up hay that I fully understood my place in this short, saturated life. I had to become a farmer. Somehow, I just had to. I had lived twenty-five years so far and seen much of the country. I had jumped off waterfalls in Tennessee and ridden a white Mustang through the Rocky Mountains. I'd driven cross-country. I'd written a book. I'd fallen in love. But I had never felt the perfect sense of knowing my place in the world till right then. Some of us are born to stoop down, touch the soil, and know it. I was one of them. I only hope I'm half as happy on my wedding day.

GETTING MY GOAT

WHEN I RETURNED TO THE SWAP MEET the following year, I left with a new addition to my flock. I didn't really mean to buy a goat. It just happened. One minute I was standing there watching the tiny herd of two-week-old lambs and kids play; the next thing I knew, I was asking the farmer who'd unloaded the trailer whether she had any wethers. She pointed to one, a small brown goat with white stripes up his face and (I swear) a sly grin. Just him, she said. I asked if she would hold him for me, and I'd be right back; I just needed to think about it. I had come to get some new laying hens and had only slightly entertained the thought of adopting a pack goat.

Just so you know, a pack goat isn't a feral goat that runs with wolves or becomes a member of your herd in some canine-inspired family bond. A pack goat is one that's trained to carry a pack. Your own personal Sherpa. The ideal animal for this type of training is a castrated male dairy goat, preferably one of the Swiss Alpine breeds. These goats have been bred for years in mountainous regions of Europe and have the sure-footedness and outback savvy to traverse even the most questionable backcountry terrain. Males, being naturally more muscular and heavy boned, make for better beasts of burden. If I could raise and train a young goat to walk on a leash, come when called, and learn to carry a small pack, I could take him hiking with me. I realized I hadn't been hiking in months.

One of the downsides of farming was the lack of time I could spend in the great outdoors. I had stopped driving to hiking trails; I was spending so much of my time in the great backyard of farming, I barely had the desire to hit the parks. I used to go hiking up into the mountains to be out in the fresh air, work my muscles, feel the soreness and the draw of nature. But farming had taken its place. I spent so much time around plants and animals (even if they were broccoli plants and chicks, not trees and wild geese) that I no longer felt that need to grab maps and a compass and head for the hills. When I did have a free Saturday and all my chores were done (a rarity), I found myself wanting to spend it *indoors*. I'd haunt a bookstore or coffee shop for a few hours, see a movie, go to a friend's house to play my guitar by the woodstove.

I realized that as much as I loved everything that made up my life, I missed hiking. I missed the way it felt to load up the station wagon at eight in the morning and leave it at the trailhead until I returned hours later, exhausted and starving. I missed how it felt to crash up a mountain, stand on a summit, and listen to the world, all tired and thin feeling. I'd pull out a tattered paperback and read the heart sutra, or *The Dharma Bums*, or a how-to guide about farming (before farming kept me leashed to the house).

But a pack goat! What a reason to step outside again! It was the perfect bridge between farm and woods. This little goat could be my passport back into the wild, a goat to run behind me on the trail, past waterfalls and chasms. We'd tramp through the woods, my solo humming punctuated by the occasional bleat from my scrappy brown pack goat. And when we stopped for me to read by the water's edge, I'd let him graze, and I'd sigh as I put my hands behind my head and look at the sky again. I missed the mountains so. A mountain goat would bring me back.

Sure, the goat seller replied, she'd hold him for me. The chances that anyone else at the fairgrounds would want a male goat were slim. The females had the advantage of offering milk and future kids for meat. But a mongrel male like this was destined to become goat curry unless some sucker decided she wanted a pet. I was that sucker. I wanted to raise a kid.

After a short walk around the festivities, and the purchase of a pair of started pullets (production reds), I returned to the goat pen. I couldn't talk myself out of it. I wanted a goat. I already had the shed, the fences, and the hay a person would need, not to mention three happy sheep who were raised with goats. And this little guy — well, he was clearly being raised with lambs, so he'd be used to the company.

For two crumpled twenties and a folded ten-dollar bill, he was mine. The goat lady put a red collar around his neck and placed him in my arms. I remember being shocked at how warm he was. He felt like a black dog that had been lying in a sunny spot all afternoon. His giant hooves hung over the crook of my arm as I hugged him. He didn't cry or fuss, just nestled his head into my chest. That was all it took. Any sense of regret or anxiety dissappeared as I buried my head in his baby hair. We walked through the festival like a mother with a new baby clutched to her chest. People stopped and stared, smiled, and took pictures. A lot of the people gave me that glance I had become very familiar with — that look of excitement and envy future farmers dart at each other when one takes a big step. I had given that look to Diana as we carried her new calf across the creek in Idaho. I had given it to my neighbor Roy when he came home with his new tractor. And a few people were giving me that same look as I held my very first goat in the folds of my beat-up jean jacket.

We got to our spot in the parking lot, and it hit me that I wasn't prepared for goat transportation. The only cage I had was full of new chickens, and I wasn't anywhere near a pet store. There were some hand-hewn wooden cages for sale, but they would cost more than he did. So I did what any sensible Vermonter would do. I set him down in the passenger seat, walked around the car to my side, and drove away. He stood up and put his hooves on the dash. Then he turned around in quick circles. When spinning grew boring, he sat like a puppy and started eating the nylon webbing of the seat belt. Whoever invented the car seat deserves a Nobel Peace Prize, I thought.

After a few minutes, though, he was fine. He was great company, actually. He folded his small front hooves under his chest, splayed out his legs, and simply rested. He looked as if riding shotgun was as normal to him as drinking milk and running around a pasture. I reached a hand over and scratched his ears as we drove past the State Line Diner. I wondered if I had broken some sort of state law by transporting livestock from New York without paperwork. I asked the goat if he knew the procedure, and he replied by closing his eyes and laying his head across the dog-hair-covered seat. I considered that pleading the Fifth.

"We're going to need a name for you, little man," I said, and tried to come up with something appropriate. He was brown and white, so I thought Tin Type might be fitting.

When I got home, I called my friend Kevin and ran it past him.

"Can I name him?" he asked.

"Whaddya got?"

"Name him after someone from *Buffy*," he said. (*Buffy the Vampire Slayer* was our favorite TV show.)

"Uhhh . . ."

"Name him Finn!"

And so Finn was the name that stuck. But it wasn't because of the famous scrappy Huckleberry of our nation's literary history. It was after Riley Finn, the military special-ops ex-boyfriend of Buffy Summers. He was my least favorite character of all time. As a nod to Kevin, Joss Whedon (creator of *Buffy*), and my pop-cultural upbringing, my first goat was named after a B-list character on a defunct television program. A little of my past life was blending with my current one.

And so I became a single parent raising a goat. I'd wake up ridiculously early and head out to the makeshift kid pen I'd designed. It was a combination of what I think was an old potting table, chicken wire, straw, and some metal roofing, but it worked fine. Inside it was an old dog crate, which I'd lined with fresh straw. Every night the little guy crawled inside to sleep.

In the morning, I'd tiptoe over to his pen just before dawn and see how close I could get before waking him. I could usually get right up to the edge of the pen, then whisper-sing "Good morning, Finn . . ." and he'd come bounding out, ears flapping and tail wagging. He'd jump around inside the pen as I set aside the bottle of milk replacement to open the gate. As soon as he saw breakfast, he'd start carrying on as if I were the most wonderful thing that ever happened to him. I'd untie the baling-wire latch and he'd nip at my fingers, nickering at me to be let out.

When he did burst free, he'd bound out and do a few laps around the chicken coop, always coming back to my side. From what I'd read about raising goats, their personalities are often compared to those of clingy dogs; a goat that was attached to a person would not gallop off into the wilderness if it knew your property was where it ate and had shelter and family. If I walked over to the garden or the sheep pen, he would come running with a grace you wouldn't expect in a goat. When he really got going, he'd snort, and I wondered if I should have named him Ripper instead.

Our morning ritual went like this: Finn would follow me around the farm while I fed chickens and sheep, weeded the garden, and carried buckets of water. As I did chores, Finn created his own series of games to keep himself interested while I pretty much ignored him. These included games such as Look How Far I Can Jump Off This Thing! and Head-Butt the Angry Rooster! Even before my first cup of coffee, I was smiling and laughing (a rarity before a goat came into my life).

When all the chores for the other livestock were done, the little guy would join me on the old wooden porch for breakfast. I'd sit in the ancient metal chair I'd outfitted with a new cushion and he'd down two bottles in my lap. It was a quiet, beautiful thing to experience before heading into the office. Feeling the heartbeat of a warm young animal as he sits in your lap, stroking his head, and giggling as his tail tickles your legs . . . it was like living in a storybook. Writing about it feels almost made up, yet it happened every day. And of all those mornings, it was the rainy ones I remember most fondly. Something about the sound of warm rain on a late-May morning, and the smell of coffee on the stove inside the cabin as I fed Finn his breakfast became a potent sensory memory. Some people never forget the feel of wet grass under their feet during their first kiss or the way the snow smelled the night their first child was born; I will never forget the feeling of a warm baby goat in my arms on a rain-soaked morning. It's not exactly romantic or epic, but it was a genuine experience and something I was starting to crave more than caffeine.

So this was what I did for those first few weeks. After my morning shower and a few cups of coffee, I'd load up Finn in the back of the Subaru in Annie's retired dog crate and head off to work. During break times and lunch, I'd let him out to pee, run around, sniff the dogs my coworkers had brought to work, and, well, be a kid. He was a hit at the office; no one seemed to mind this weird breed of sporting dog that had taken up residence on the back lawn.

In the evening, after the chores were done and the animals fed and dogs walked, I'd grab Finn's lead and we'd go for a walk, even if there were only twenty minutes of daylight left. We didn't walk far; I usually had a stomach full of food and would be growing tired. We moved slowly, a postprandial jaunt over the little dirt bridge that spanned the stream. We'd head down to the main road, and every now and then Finn would stop to nibble a dead leaf on the ground.

We usually didn't see a single car. I'd listen to the sounds of weather changing — leaves tossing in the limbs above us, a burning brush pile crackling on someone's property. The air smelled like smoke and cut grass. We'd stop in the cool shade of a sugar maple and the wind would rush warm air into us. Finn, confused by the sudden change in the world, would bow down on his front legs and then jump into the air, throwing his horns into nothing to fight the barometrics.

If I'm lucky and get to live into old age, I'll look back on these rituals and be glad. I'll remember the summer nights at the cabin walking silently alongside my young goat, scanning the trees for fireflies. Perhaps, as an old woman, I will find myself on a walking tour of someone else's small farm on a rainy morning. Because I am me, there will still be a mug of coffee in my thin hands, and when I approach the goat pen and smell that combination of goat hair, straw, and hot coffee, I will almost fall to my knees from the nostalgia. Raising Finn was turning out to be a revelation.

YOU NEED A TRUCK, GIRL

FINN JUMPED OUT OF THE DOG CRATE in the back of my Subaru as if tailgate leaping were a caprine Olympic event. With his tail high, front legs tucked under his chest, and back arched, he looked more like one of the cardboard reindeer decorations of my childhood than the son of Cain that he was. He landed on the grass, twirled around, tail wagging, and bleated for some milk from the bottle in my hand. I beamed at the little guy and fed him, as some coworkers strolled out to enjoy the sunshine. We were parked only a few feet from the herd of motorcycles the writing staff drove in on, without fail, every sunny day.

Paul, Eric, and Tim started up some small talk. Finn splayed his legs and peed, having downed two bottles of milk in a row. The guys laughed and made fitting comments. I joined in. There was no point in trying to class up a peeing goat.

While we enjoyed the sun and talked about work, pets, and the weather, Eric crossed his arms and leaned back onto the heels of his shoes. He smiled and shook his head as he looked at the back of my Forester. I hadn't even thought about shutting the hatch of the car, but as Eric looked into the back of it, I was able to see it through the eyes of a normal non-goat-owning person. It was a horror. The once pristine car was full of dented dog crates, hay, feed bags, old quilts, garden flats, mud, and dust; what little room was left was occupied by egg cartons. It was the aftermath of a very intentional life in an unintentional vehicle.

My car had become the victim of homesteading. It smelled like dead grass and wet dog. The windows were smudged with nose smears and the sides ruined with claw marks. "You're never going to be able to sell that car," was all Eric had to say through a smirk. He was right. And the more my life careened from part-time homesteader to full-time farmer, the more I was realizing I'd better start thinking about my automotive future.

I needed a truck. It was time. I'd been hauling hay, tools, feed, and every farm critter imaginable in the back of my eight-year-old station wagon. Since I'd started down this feed-sack-lined path, I had made passengers out of boxed honeybees, chicks, turkey poults, ducklings, goslings, bunnies, chickens, geese, fiber rabbits, roosters, three sheep, two dogs, and a goat kid. I had transported vegetables, topsoil, hoses, tools, Havahart traps (occasionally inhabited), dead chickens, and everything else that comes with living on a small farm.

While the car had served me well, there was no question it had suffered. Carpeted backseats were not meant to handle a flock of sheep; there was still a stain from when Maude peed all over it (I didn't realize the tarp I'd put down had a hole in it). Clearly, it was time for me to start looking into something meant to take such a beating. A pony is not a draft horse, no matter how many times you put the hames over its head.

Eric wasn't the only one who thought my car days were behind me. It wasn't uncommon for me to inspire jeers in the parking lot of places like Home Depot and Tractor Supply. My station wagon — in all its topsoil-and-dog-hair-encrusted glory — would pull up to a line of trucks and park between them like a sparrow among crows on a telephone wire. When I'd open the hatch and start piling in bags of compost or feed, kicking up clouds of dirt and bits of hay, someone would inevitably get into the truck next to me, shake his head, and say, "You need this truck more than I do, darling."

Money was tight, though. Another car payment was out of the question, so if I wanted to get a truck, I knew it would be used, for cash. My budget would not allow spending a lot, either. My first searches at preowned-car dealerships and the weekly flyers seemed to suggest that to find something serviceable, I needed to have at least three times what I'd hoped to spend. The Internet turned up cheaper options, but they were sketchy at best. On Craigslist, folks were selling trucks for two hundred dollars that sounded more like a death wish than a farm helper. I'd come across ads that said things like, "Bottom's all rusted through but I got plywood down there and new brakes last spring. She'll pass inspection if you know someone." Great.

Sheer luck had landed me a respectable sum in the shape of a book advance, and it was enough for me to work with. Hopefully, I could find a small pickup in fairly decent shape for my budget of three thousand dollars. I had two caveats: It had to be automatic and it had to play music. I knew myself well enough to realize that I would be driving with a hot mug of coffee in one hand, a dog in the front seat, CDs changing in the stereo, and a ruckus in the back bed, depending on the season and the livestock being moved. This is not the lifestyle of someone who needs to concentrate on a stick shift.

Every morning I read the Craigslist postings, hoping that some family somewhere had brought home twins and needed to trade up to a minivan. I figured somewhere in Vermont a small single cab had to be waiting for me.

Turns out there wasn't.

I finally gave up looking in the Vermont classifieds and started looking in New York instead. After all, Washington County had scads of trucks on its county roads, far more than the Subaru-speckled country lanes over on the Vermont side of the border. A few days into my search, I came across an ad for White Creek Auto, just a hop over the state line. An ad for an automatic side-step 1999 Ford Ranger came up. Her paint was called Copper and looked like a burnt orange-red. She must have belonged to someone who never farmed a day in his life (God bless him) because, from the photos at least, she looked pristine. She was beautiful.

White Creek Auto had it posted for twenty-nine hundred dollars, with over a hundred thousand miles already on the odometer. While on the phone, the manager explained that they'd gotten her at auction dirt cheap and were selling her as is. I clicked over to the White Creek Auto website and read its slogan, choking on my Saturday-morning coffee as I did so: **WHITE CREEK AUTO: HOME OF THE ABSOLUTELY NO GUARANTEE WHATSOEVER!** Well. I guess that covers it.

I told him I'd be down in a few hours to take it for a test-drive.

My friends Phil and Mike came with me to see the old truck. Both lectured me about the art of bargaining, that I should lowball the guy and expect to walk away. They were better at this sort of thing than I was. Honestly, if this truck was in good shape, ran well, and had a sweet stereo system, I wasn't about to turn a 180 and stroll into the sunset for a lousy two hundred bucks (of course I considered far more important things, like four-wheel drive, mileage, gas consumption, and amount of rust on the body, but somehow, they didn't seem as important to me as a decent stereo). I could sell my banjo and make two hundred dollars. If this animal was sound, she was going home with me.

When we pulled into the garage's driveway, I noticed the little pickup instantly; she looked as if she had just come off the showroom lot. I speed-walked over to it and opened a door to check the interior. There were a few cigarette burns here and there, but it didn't smell like an ashtray. Inside the small khaki cab were two giant coffee-cup holders, a CD player with radio, a few storage compartments, and CD slits in the visor. Perfect.

The owner of the garage came out and asked if I was the girl who e-mailed about the Ford. When he saw how excited I was, he knew he had a sucker in his claws and handed me the keys and a magnetic plate. Mike, ever the watchful big-brother figure, agreed to come along on the test run. He drove a new Toyota Tacoma and loved it. I had ridden along in his pickup and loved it too, but I also knew such a vehicle cost more than a down payment on a farm. And I wasn't looking for a shiny new commuter like his, anyway. I needed a workhorse.

We climbed into the cab, and I took a few seconds to get my bearings. I turned the key, felt the engine rev, and hit the gas. I don't know if Mike noticed the grin on my face, but I was elated. My feet almost floated off the pedals as we turned the corner down the dusty road.

We drove a mile or so, and Mike suggested we pull her over and look under the hood while she ran. We both felt she passed the driving and brake-slamming acrobatics we had put her through — now we wanted a physical to prove we were right. Leaving the engine idling, we popped the hood. All looked well inside.

It wasn't long after we pulled her over that another truck started down the road and saw one of its countrymen in "distress." A gutted, windowless, 1980s Toyota with a flat wooden bed pulled up with three people jammed inside: the driver — a wiry guy in his mid-thirties — and two silent, equally skinny teenage boys I suspected were his progeny.

"You okay?" he asked, sincerely.

"Yup! Just taking this used truck on a test-drive, checking what's under the hood."

"Whaddretheyaskinforrit?"

"A little south of three grand."

"Really?"

"Yes."

"Buy it. This heap of crap cost me three grand. That truck looks brand new."

Feeling like I had received a benediction, I drove her back to the garage content, thrilled even. Mike smiled, too. We had both made a new friend in this decade-old orange truck.

After an intense sparring match of bargaining with the owner, I sealed the deal for the unbeatable price of twenty-nine hundred dollars — exactly what the ad requested. He simply would not budge on the price. He said a truck like that didn't sit on his lot long, and I believed him. He wasn't feeding me a line as much as he was simply making a statement. So we sat down, and I wrote a check (the largest check I had ever handed anyone), and within moments the title information was handed over and the paperwork mailed to Albany. He handed me the key, shook my hand, and told me to take care of her. Barely believing this day had come, I took the small black key from his hand and waved to my friends, then we hopped into our separate rides to head home. I pulled a small black CD case out of my shoulder bag and popped a disc into the truck's CD player. Old Crow Medicine

Show's "Wagon Wheel" blared through the speakers, and as the fiddle started to croon into the first verse I pulled out of the driveway. This greenhorn farmer had finally entered the world of cabs, beds, and tailgates.

Despite a quick return to the garage to have a wheel bearing replaced, the thing ran like a thoroughbred. It passed Vermont state inspection with flying colors, and I was secretly elated when the garage gave me a 10 sticker instead of a 9. It was late enough in the month of September for the inspection to count as October, so my favorite month, my holy month, was birth-marked on the windshield. I considered it a good omen.

The truck became my main vehicle. I took it everywhere: to work, on weekend errands, for dinner in town with friends. But this was also a working vehicle. It was constantly loaded with hay or straw, feed, and fencing. I drove it to pick up fifteen chickens with a coworker, and we happily carried our stock home in the bed. As autumn turned deeper into her burning days, I started the simple ritual of driving it a few miles down the road from the office to Clear Brook Farm, where I could buy freshly baked breads, cider, and cheddar cheese. I could have a baguette with fresh curd and wash it down with sweet, freshly pressed cider. I would come back to the office, sit in the bed with my picnic and a book, and savor my lunch break with a new level of gusto.

When your life is one of gardens and livestock, a pickup truck becomes an extension of that life. It didn't take long for me to understand the affection of a hundred country ballads or the gruff conversations going on at the back of the Wayside. I used to roll my eyes and smile as I heard those men talk about their trucks with such passion, concern, and protectiveness. If someone needed to buy a new one, watch out. The debates over makes, years, and models would be loud enough to spill your coffee. But after driving one, depending on one, and learning about the flexibility and freedom the thing granted me so easily, I was hooked — a convert to the First Church of the F150, Our Lady of Silverado.

BUILDING PARADISE BROKE AND ALONE

WHEN AN OLD TRUCK IS CONSIDERED a luxury purchase, you can safely assume that I'm not a wealthy individual. I've lived, and continue to live, paycheck to paycheck. It's not something I'm particularly proud of, as I've never been much good at saving, but I am proud of the quality of life I live when I'm not freaking out about money.

My first year in Vermont, things were especially tight. The first paycheck of the month went toward rent, the Subaru payment, food, and utilities. The second was eaten up by bills, credit cards, and things like gas and dog food. What was left over was usually around two hundred bucks; that's what I ran my backyard farm on. Sometimes that was more than enough, and sometimes it barely got us through. When times called for drastic measures, I had to start selling things to make ends meet. My first winter with sheep, I sold my beloved banjo so that I could buy enough hay to get us through till spring. I foolishly budgeted for the fencing, the shed, and all the hardware that goes into the work and then realized the sheep probably wanted to eat at some point. I played one last sad song on the banjo and sold it to a student at Bennington College.

Now, two hundred bucks to manage a small farm for a month is plenty, if all you're doing is feeding a small flock of chickens (I'd say you'd have about $175 extra, quite honestly). Taking on even just three sheep meant a whole new world of start-up and sustaining costs. I needed money for such things as hay, fence repair, shelter, winter water-tank defrosters, hoof trimmers, Pro-Pen antibiotics, and my very own assortment of syringes for livestock injections (something I never thought I'd have in my medicine cabinet, short of developing diabetes or a heroin addiction, both of which I hoped to avoid).

Expanding the garden was expensive, too. I didn't have a tiller to maintain, but I quickly learned that I needed to invest in decent hand tools. After the cash was spent on seeds, tools, and a few six-packs of veggies, I realized I'd also need to expand the garden fencing, put in a scarecrow or plastic owl, and buy netting to keep the birds out of the berries. Most of these were start-up costs, but as a renter, every new backyard is a start-up. I was getting tired of breaking all that fresh sod, too.

This was the reality of my new life. Homesteading on a larger scale simply cost more money. But there was some income slowly filtering in as well, which is more than many folks can say about their backyards. At the office I was selling eggs for a couple of dollars a dozen, and the occasional apple pie, too.

I had learned to breed my French Angora rabbits successfully, and their pedigreed offspring could fetch up to fifty dollars a kit. The chickens were buying their own feed and the occasional cup of coffee for their farmer, and the rabbits were certainly earning their annual room and board with the sale of woolly bunnies. I taught beginner fiddle and dulcimer workshops. And whenever things got really tight, I'd fall back on selling things on eBay and eating a lot more homemade food.

Homesteading saved money, too. I was eating really well from my giant backyard garden. My second growing season at the cabin, I had fifteen raised beds, each with about twenty square feet of growing space. I had corn growing taller than I was, pumpkins curling around the fence posts, and tomatoes so plump and red, I was worried they'd burst open if you stared at them too long. I canned and froze my garden harvest and baked my own bread and pizza crusts. Weekends, I ate entirely from my little homestead, with breakfasts of garden omelets or homemade French toast or pancakes, lunches of salads and cold mint tea, and dinners of savory pasta, baked squash, or bubbling pizza with my own handmade mozzarella.

By my second summer at the cabin, eating out of the garden had saved me enough money that I could not only afford to keep the farm afloat, but I could also buy a new banjo to pick in the hammock I'd strung up between two ancient oaks. Some weekends I would park the car in the driveway and not get into it again until Monday morning rolled around and I was off to work. I can't say I was such an honorable pioneer all the time, but most of the time I was content with a blanket, a book, and a new song to practice on my violin in the sheep pasture. I'd sprawl out there on a blanket for hours with a cold drink and some instruments, and eventually Sal would saunter over for an ear scratch and lie down beside me. If I could manufacture the sense of home and peace I felt lying on my back in the sun, among the tall grass, sipping some hard cider, listening to the quiet cud-chewing of my flock as my fingers mindlessly plucked away at my strings, I would be wealthier than Martha Stewart.

To keep the farm running, I had to plan for a string of summer weeks that required practically no money at all to prepare for what I called the Fall Wall — that time from September to the first snowfall when suddenly life in the Northeast gets very pricey: firewood, heating oil, winter hay (I now stored it in the garage to keep it dry and parked my car outside in the driveway), snow tires, and warm clothes. I rarely shopped, and when I did, it was in secondhand stores as much as possible or at sales at places like the feed store. I used to blush when people asked me where my sweater came from and I had to say "Oh, clearance at Tractor Supply: five dollars!" but I got over that quick. I might have been broke 99 percent of the time, but my money was being funneled into my experience and living the life I'd always wanted to live. So what if everyone else from my college design classes were taking weekend trips to the city or filling their homes with furniture from Crate and Barrel? I was taking weekend trips to buy livestock or see local sheepdog trials, and I actually *used* crates and barrels every day to store tools and chicken feed.

Some folks might say it's careless, dangerous, and certainly foolish to start farming without a respectable savings account and more experience. But quitting my day job and going back to school for agriculture or taking up a resident internship at an organic farm wasn't possible. I had student loans to pay off. I also had real-life bills, from living on my own since college. No, deciding to wash my hands of the nine-to-five world and living off my backyard was not an option, not unless I wanted to lose my health insurance, declare bankruptcy, and have the dealership take away my Subaru.

So my options seemed pretty cut and dried. I could save money and wait until I was closer to retirement age and buy a small farm responsibly. Or I could just jump into it, become a student of Hard Knocks University, Agricultural Department. I chose the latter. It suited me to dive into the life I wanted. Also, something I read along the way really stuck with me. In Joel Salatin's book *You Can Farm,* he encourages the hopeful farmer not to put off her dream until the circumstances are perfect. He says, quite bluntly, that if you can't make a profit on your backyard or garage, well, what makes you think you'll be successful when you finally do get hold of some land? His point was to just get started. Learning comes from mistakes, experience, and hard work. Ideally, I would have grown up as a third-generation farmer who already understood how to clean a roasting chicken and raise heirloom vegetables, but that wasn't me. If I wanted this to happen, I had to start now. Perhaps then, when the land and life I dreamed of did come along, I would already know a few things, and getting started in the farming business would be more about growth and transition than about making the same mistakes with electric fences and gardens I would have made for years on rented land.

The reality of this gung-ho attitude isn't as sexy as it sounds, of course. It was fraught with panic attacks and the feeling of treading water. What had me pounding with anxiety wasn't being tight on cash; it was the fear that there would be no cash at all. When the economy came crashing down and people were being laid off at the office, I was a wreck. Losing my job was all it would take to end my homesteading education. Some nights I would lie on the futon in front of the fire, unable to sleep, because I knew how close I was to the edge. If I lost my job, the dogs and I would have to move to some apartment and try to get another job while I got back on my feet. The chickens would be sold. The sheep would be returned to their former farm, and rabbits and bees would end up as Craigslist castaways. This is a sobering thing to think about when you are planning gardens, breeding livestock, and building new animal housing. A signature on a pink slip and a handshake could make everything go away. By morning these thoughts would have subsided, and I'd be right out there again, planting and writing rabbit pedigrees, but the haunting stayed with me. Homesteaders with mortgages must sleep better than I did, I figured.

You might also say that homesteaders with *partners* sleep better, too. Not that I would know.

I'm not going to lie: the pickings in southern Vermont were slim. After living in the area for a year or two, I'd only managed to scrounge a handful of dates, none of them good. This wasn't their fault; I was horrible at dating. I would get so nervous about the first impression I was making (you know a date isn't going to end well when you're all wound up, self-conscious, and scared your mascara is leaking into the dark circles under your eyes). I couldn't relax. Instead of being the confident woman from the online profile I'd created (one of the only ways to meet people in my area code), I came across as a distracted, fast-talking hurricane, barely able to understand the answers coming out of my own mouth, much less able to digest his. I was a total disaster.

The other major dating issue was less a problem of quality as it was of quantity. There were handsome, caring, and intelligent men out there on the backs of tractors and quarter horses, but when compared to the sheer quantity of single women, it became obvious that there was a serious John Dear deficit. Besides, the few men weren't that easy to meet, unless you had the timing skills of a stalker and were willing to haunt the local Agway. Clearly, I wasn't up on my creepy skills, but I had high hopes. Eventually, I'd stumble across some guy who wanted a woman who enjoyed *The Daily Show* as much as she enjoyed milking dairy goats.

Being a single woman in a rural area also seemed to limit my options to two main categories of men. The first was the backwoods cliché and the second was an awkward grab bag of outsiders.

Now, I appreciated and couldn't help but be attracted to some of the Good Ol' Boys on farms, in construction jobs, and working in auto shops. These were the guys who unapologetically stocked their closets with their best suit and tie, running sneakers, and bottles of deer urine. They worked hard, played hard, and enjoyed the aspects of rural living that make them the butt of urban jokes. They hunted, rode four-wheelers, drank Bud, and knew where their grandparents were buried. The good ones were clever, kind, fiercely loyal to their families and faith, and generous with what they had to give, as long as they didn't feel taken advantage of.

Unfortunately, all these men were married. Moving on.

The second category was the kind of man I kept running into: gentlemen outsiders. These were either locals who had lived their whole lives in the area but never felt accepted or guys who, like me, had moved here, away from their roots, and were trying to start a life in a strange place with uneven footing. For whatever reason, they took a job, or followed a dream, or by some strange set of circumstances discovered themselves in the land of milk and honey (we had a lot of dairy farms and beekeepers) and weren't quite sure what to do next. They were always passionate about one thing, be it music, or art, or animals, or their career, or whatever it was that kept their hands from being idle and put a spark in their eyes. I was a sucker for these guys. I went on dates with men who did sound checks at rock concerts, wrote screenplays, and played banjo in the apartment they shared with four roommates. The guy with a record collection and a smirk will always win the day over the guy who actually knows how to start a tractor and isn't scared of horses.

In short, I was an idiot when it came to love.

I'd never had any luck in the love department. I'd only really loved two men in my life, and neither of them felt the same way about me — not even close. The first time was hard; the second time, debilitating. I found that it got harder as I got older, and not because of clichés like "all the good ones are taken," but because I was growing to know myself better, and with that understanding of what I wanted for myself I had a pretty detailed idea of the kind of man who could fit into that life. I held this ideal in my head, and anyone short of it didn't get even a wave. Unfortunately, this limited the dating pool to such a thin sampling of the population that it was rare for me to find myself having coffee with someone in a bookstore even.

I also confused friendship with romance all too often. Part of me wanted to blame the fact that I was raised watching *Cheers*, *When Harry Met Sally*, and *Say Anything*. The other part blamed myself. I'd get the possibility of a relationship in my head and fall in love with that possibility before the guy knew what was happening, and suddenly going for a late-night walk to get a cup of coffee was saturated nostalgia for me and only a way to spend $1.89 for him.

I hadn't given up on falling in love, but the pragmatist in me had trouble believing it would happen. I wasn't sure I'd ever feel the same way about a man as I did about my farm. Ever since I'd held that first chick in my palm or slammed that first hoe into the soil in Idaho, owning my own farm someday had felt possible and real. Every atom in my body believed I could get there. My faith that I would someday walk out onto my own pasture with a sheepdog at my side wasn't something

I had to question, because farming was always the end result. I had no plan, no money, and no experience, but I knew I loved it, and I could take that blend of hope and willpower and sign those mortgage papers when the time came.

You see, you can love a farm and spend a lifetime dedicated to it and know that it doesn't have to love you back. I knew my farm would sustain me and heal me and occupy my body and mind in a way that was not unlike love. Farming is physical and emotional, and you get out of it exactly what you're willing to put into it. Like any other relationship, it has its high and low points, but the deep undertone of pure, unadulterated love keeps it pumping blood. It's something you can have if you're willing to work for it.

Sadly, it's not easy to find people who are interested in this sort of life. A guy who still enjoys eating chicken after slaughtering thirty that same morning and isn't tired of weeding in August is a rare find.

I have no idea who this guy is. I just hope he has a decent record collection, likes dogs, and has good timing. In my experience, one of these is always lacking.

SAVING SARAH

ONE BRISK FALL MORNING I found myself standing in the sheep pen with a fiberglass staff, yelling commands at a young sheepdog. I stood there under the red maple, watching my confused sheep circle in panic from the little black-and-brown dog turning them around the pen. I suddenly realized how absurd the training session was. I had only been involved in the world of border collies for a few months, after all. I'd been diligent, sure: attending trials, training sessions, and lectures. I'd read books, visited farms, helped at novice practice sessions — but now I was standing in a pen with my own sheep and my own sheepdog, trying to remember how to go about this.

In the haze of the confusion, though, there was also real joy. Ever since I'd first learned to knit, I had secretly wanted my own fiber animals. When I finally took the plunge and started homesteading, I bought, raised, bred, sheared, spun, and knit from my own small herd of Angora rabbits. But sheep are the ultimate homesteading beasts. You can eat a sheep, milk a sheep, or wear a sheep. This was an animal that made you everything from roast, to ricotta, to rag-wool socks.

The more I learned about them, the more I made them a part of my life. I stopped buying polar fleece and switched to wool instead. My dogs ate lamb-based dog food. I went to restaurants with lamb on the menu; bought sheep cheese, fleeces to spin, yarns, and books. When I started to eat meat again — after almost a decade of vegetarianism — it was pastured lamb that first passed my lips. I was becoming a hardcore advocate of the ovine marketplace. I did this because I wanted to support the product I would be producing on my own land someday. Helping the sheep industry was helping myself, along with all the other shepherds who were trying to make a go of it.

Deciding to dedicate myself to the noble hogget meant I had to face some pretty harsh realities, though:

1. I couldn't run a sheep farm by myself.
2. I'm single.
3. I needed a border collie.

I could be a shepherd if I only had some help. A good, dependable working dog was just what I needed — an animal that moved sheep, kept them at bay when I was pouring grain, separated sick ewes from the flock so that I could administer medications, watched and protected the herd. Having a sheepdog was about so much more than herding sheep; it meant the freedom to farm as a single woman. Yes, I only had three bum sheep and a rented backyard, but getting a young dog trained and started would mean that when I was ready to buy a farm and start working my own flock, I would be prepared. I needed this dog to become the farmer I wanted to be.

And so there was Sarah, a started dog versed in basic herding. She was around a year old, and already herding dozens of sheep a day on a small farm in upstate New York. A member and trainer I had met through NEBCA had been working with her. She had come to the trainer's farm as a reject. An elderly gentleman had bought her as a puppy from a working-dog kennel, thinking they would live out their days together on some rural property upstate. But he became ill and had to move back with family near the city. Now this farm pup was being raised in Manhattan — total sensory overload for a dog bred to focus on one intense job at a time. Predictably, the confined spaces, high stress, and family life proved too much for the little girl. The family sent her away to a shelter: sadly, the fate of many pet border collies whose owners think they're getting a cute and cuddly family dog but end up with a puppy that circles their children's playgroup like a land shark. Border collies, especially ones bred from strong herding lines, are not golden retrievers.

So Sarah was sent to a shelter, discovered by local border collie–rescue activists, then sent to Barb and Bernie Armata's farm, which is where I met her. She was in her glory, smiling and panting, working a large herd of Scottish blackface sheep every day. Her character review was stunning, her demeanor calm; she was a love bug and already tuned to the work I needed to be done.

I felt prepared to take this step. For months I had been immersed in this world. The trainers and club members knew me and understood my tenacity and willingness to dive headfirst into this lifestyle. So many of them had once been in the same position, just starting out with a few sheep and looking for a working dog. So when I put out the word to the border collie community that I was looking for a "started" dog (meaning a young dog with some professional training under her belt — in other words, a puppy I couldn't screw up), I heard back from Barb and Bernie. They thought that this ex-city-livin' farm girl might be a good match for their ex-city-livin' farm dog. We chatted over e-mail and sent photos back and forth. Sarah was a tricolor rough coat and small — only about thirty-five pounds. But like most working border collies, she wasn't bred to meet a physical breed standard; she was bred to herd sheep. She was floppy-eared and scrappy, and her coat, while technically rough, was only an inch or two long. She was the picture of a farm dog to me, though, and had that bit of wildness in her eyes. So I made arrangements to meet up with Barb and Bernie at the Fall Foliage Sheepdog Trial in Cooperstown, New York. I could hardly wait to meet the dog in the photographs.

I had friends visiting from Pennsylvania the weekend of the trial. They were coming up to Vermont during peak foliage time to see the farm, have a small vacation, play some music (they're both fiddlers as well), and join me for my three-hour pilgrimage to Border Collie Mecca. I explained to them that this was one of the most prestigious trials in the Northeast, and people would be competing from across the East Coast and Canada. They were sweet and polite about it, and more than willing to tag along, but clearly my enthusiasm wasn't contagious. They did know about the potential farm dog, though, and were as excited to meet her as I was.

So the three of us drove south to Cooperstown, with Jazz and Annie along for the ride. Our little caravan of dogs and ladies rolled through the autumn glory of the Hudson Valley, and we found ourselves being bowled over by it. The entire Northeast was at its most colorful, but traveling only three hours south, I could see how much louder the color became. Vermont was looking good, but New York was putting on a fireworks display. The sight of it lifted my spirits even higher.

We arrived at the trial fields around ten in the morning and set up our camp of folding chairs by the gate near the starting post. That sounds more exciting than it is, actually. The starting post is not some latched series of gates that release a pack of sheepdogs, like at the Kentucky Derby. It's simply a white post, maybe five feet high. It marked the starting point for all dogs being put through their paces. While my friends went about setting up their chairs and backpacks for optimal sun and comfort, I was scanning the crowds for Barb and Sarah. I could not wait to meet this dog.

I finally found Barb talking with some other trainers near the main tent. We shook hands, and Barb told me that Sarah was over near her car and went to get her. When she returned, it was with a long thin lead of braided leather attached to a little sprite wagging her tail. Barb handed me the lead, and I knelt down to meet Sarah. She paid some attention to me but kept darting her head back to the right. She was fixated on the sheep. At the time I took this as a great sign, but in retrospect I should have understood the warning. She was intense. She was a working dog through and through. New people, a crowd, cars, trucks, hot-dog stands . . . she didn't care. Sarah wanted one thing in this whole world, and it was covered in wool on the other side of the fence. I had signed up for a herding dog, and this was it.

We spent the trial together, and for the first time in all my herding club adventures, I had a dog with me. I saw a well-known trainer I had spoken with at a few events. He knew I was hurting for a dog, and bad. When he came upon the happy couple of girls smiling in the early-afternoon sun, he called over, "Jenna, you got your dog, huh?" We parted, and as I walked away with the little black dog by my side, I felt different, better. Having a few sheep in the backyard was one thing. Having a sheepdog meant I was serious about this sheepherding business.

A few weekends later I drove back to Barb and Bernie's farm to pick up Sarah for good. It was a rainy day. It was also the open house for Barb's new venture, a top-of-the-line boarding facility for dogs. Barb was living the dream: she was finding a way to make shepherding on her farm a full-time job. Bringing the business of people and their pets to her home would provide a paycheck and the peace of mind of being near the flock.

Sarah and I stood together in the corner of her office, among the crowd of people there to see the kennels. Barb showed me where to sign the papers, then took us out to a small training field with a few quiet sheep. She handed me the staff and showed me how to use it to signal to Sarah and keep the sheep between us. I was walking backward while the flock charged toward me, with Sarah weaving and darting on the far side of the ewes. The lesson was short but good. I understood my job, mostly, and Sarah seemed to know what was going on. (I realized any sheepherding lessons would be more for me than for her.)

As we were getting ready to drive back to Sandgate, she sat calmly at the end of her leash and seemed to catalog everything going on around her. After some well-wishing and waves good-bye, I pulled out of the driveway as a shepherd with her collie. Sarah was in a crate in the back of the station wagon. I was unsure of how she'd be in the car (I was told she wasn't fond of them), and to avoid the risk of a freak-out on the New York Thruway, I kept her boxed.

Once home at the cabin, she made herself comfortable on the couch with me, and we ended our exciting day by watching a movie rented from the NEBCA library, *A Year of the Working Sheepdog*. It was a documentary about a working farm in North Devon and the dogs that kept the show running. I watched it, amazed. The shepherd in the film, David Kennard, was a true farmer, a pro. I was a kid who'd just gotten her first tricycle and was watching the Tour de France. I sighed, happy. In the morning we'd work those three sheep in the backyard like Barb had showed us. But that first night was just about scratching ears and letting Sarah know she was home. She slept well and deep, and I had my collie. We had both come quite far. Devon would come another day.

The next morning, and the mornings thereafter, were not anything like the documentary about well-trained dogs on open moors. My three sheep loathed and reacted violently to being herded. I had been warned that the only sheep to get should be dog-broke, meaning "used to working with a border collie." Mine were more interested in breaking dogs. Maude stamped her feet, then sprinted into the fences in a panic. Marvin glared at the little black dog as though she'd just killed his fiancée. Sal looked at us, then the fence, then us again and hopped the fence like a friggin' white-tailed deer. He happily munched grass on the opposite side of the pen while Maude and Marvin dealt with the wolf.

Eventually, we'd get to a place where Sarah was moving them and I was on the opposite side of Sarah. We'd do this maybe fifteen minutes a morning in our small training space. I was constantly worried she'd get hurt, but Sarah didn't seem to mind the fuss. When the sheep and I were panting, I'd scoop her up and say "That'll do," and we'd go inside. Those short sessions were never enough for the young dog. She'd bark at the door while I was in the shower or making dinner.

She needed to be outside more than Jazz and Annie and I were used to.

For a few weeks we kept up the training outside, until one morning Marvin taught me a lesson I still remember, and limp on, to this day. Sarah and I were working in the pen, moving the sheep and trying to balance the flock between us, when, somehow, Sarah found herself in the corner of the pen near the gate. Marvin saw the bane of his existence trapped and acted fast. I was three steps away from the scene and rushed to break up what I knew was about to happen. Marvin reared up, lowered his massive skull, and was about to head-butt the small dog with all the force he could muster, ending the reign of terror NOW. Seeing what could be the death of my new dog, I threw my body between them and felt the force of the blow hit the right side of my right knee with a bullet of pain and a crack so loud I screamed. It was broken, I thought. It had to be. Using what wits I had left, I screamed at Marvin to back up and scooped up Sarah, hobbled out of the pen, slammed the gate behind us, and blacked out.

I woke up with Sarah sitting next to me, unconcerned, eating a stick. I tried to move and couldn't. I started crying. All I could think was "Everything my mother ever told me was right." I was taking on too much. I shouldn't be alone out here. I would get hurt if I wasn't careful. I should be married. I should be dating. I should wear lipstick. . . . I army-crawled back to the house and realized that work started in half an hour. I called my boss, wailing into the phone about God knows what, and (I think) explained I'd be late.

It took a few weeks of walking with a cane to recover from the torn muscle. And a few months for me to actually bend and move my knee like a normal person. But it did heal. And while it did, neither Sarah nor I dared herd my anarchist sheep. I didn't know what to do now. I had the wrong sheep. Should I sell them and get new ones? Should I add two old sheep-dogged ewes and buy more hay? Would my landlord allow that? Did she even know about Sarah? I hadn't told her, but I worried my neighbor might.

As her one outlet for spending energy disappeared, Sarah became nervous. Without sheep to herd, she had no purpose. She was once again a skittish pet without a job. She started nipping at me when I walked across the room, occasionally really biting into me. I'd yelp and correct her but blamed myself for the poor behavior. After all, if I had proper sheep, I wouldn't be limping around a bored and nervous dog.

My heels were not the only things Sarah bit. She snapped at a friend's kid who ran past her swiftly and then at the father when he did the same. They were both very calm about it, but I could tell it was more than a love nip. Then one day at a bookstore in town, she went too far and bit the leg of a passing staffperson. I had been bringing her into the bookstore for weeks, using the dog-friendly place as a training and socialization area, and she had never acted like this before. The person whirled around and screamed at me in front of everyone in the packed store, using many choice words and threatening to sue. I broke down in tears, apologizing, terrified of an actual lawsuit. I worked it out with the store manager but was never allowed to bring Sarah anywhere near the place ever again.

Then the final and most serious infraction occurred, at my parents' house over Thanksgiving. Sarah ran across the kitchen floor and bit my father in the leg. My dad cried out. I once again scooped her up and apologized. My father demanded that she stay in my bedroom for the rest of the visit. We had young children coming the following day and couldn't risk one getting hurt. My stomach felt as if it was filled with concrete, and panic began to set in. I knew what I had to do. I had to take Sarah back to Barb. I couldn't contain her, and now she had hurt four people. She could no longer be my dog.

Crying the entire ride, I drove the four of us north. Jazz was asleep in the backseat, Annie was curled up in the passenger seat as usual, and Sarah was in her crate in the back. I pulled up to Barb's farm and instantly lost it when I saw her. I knew, and she knew, that I had messed up. The dog needed work, and a real flock, and someone with the time and patience to train her and show her how to live in the world. I was not the person Barb or I had thought I could be. I knew handing back that dog was giving up on an animal. I felt like I'd failed the club, the trainer, and my destiny. Barb assured me that Sarah would be fine and would be placed on a working farm without children; I didn't have to worry about her. But I did and couldn't stop crying. Relinquishing a dog, for any reason, is admitting you've made a mistake that you aren't able to fix. Now she'd have to find yet another home.

I thought I was going to be her savior, and she would be my shepherd. I thought we'd be at the post, side by side, at sheepdog trials. In some way I thought she could train me to the work that seemed so right. Instead, I'd given her eight hours of solitude each day, three crummy sheep, a near-death experience, and two apathetic canine roommates. I'd failed her.

I drove home to the cabin with just two dogs. When I got back to the farm, I put the sheepdog videos in the cupboard. I was not who I thought I could be. Not yet.

TURKEY DRAMA

ANYONE WHO DOESN'T BELIEVE that birds evolved from dinosaurs has never raised a turkey. A few months into my first Vermont summer, I felt like a curator at the Museum of Natural History. The little poult I'd picked up from the feed store in May had grown into the giant of the henhouse. When the dog days of late August took over and the streambeds were bursting with high ferns and tall grass, I watched him prowl. He was a hefty, free-ranging white bird weaving through the bush like a velociraptor on the hunt. It's a happy (and rare) sight, this stalking. My boy happened to be the most popular breed of farmed turkey in America — a Broad-breasted White, the same kind you buy at the supermarket. His brethren in industrial production sites all over America would never know what a day among ferns was like. But this boy, he was living large, and it made me beam.

Some evenings I would be relaxing in my hammock in a happy place with a banjo in my arms and some hard cider at arm's length, and as I was about to close my eyes around the B section of a waltz, I would hear that mighty roar right behind me. *GOOBBBBLLEECOBOGBGOBGGOBLLLlllssshhGlg!* I'd just about fall off the ropes. Verbal sneak attacks like this were common, and eventually I learned to tell when he was coming. My ears had never been that attuned to ruffling feathers before. If someone dared to wear a feather boa, she'd better watch out; I'd spin around on her so fast that we'd both get whiplash.

I got used to the Butterball-alternative life. Every morning TD (Thanksgiving Dinner) met me with gobbles and fantail shows. I watched him follow the geese and ducks all over their small paradise. Often, he would wander down the dirt road to my neighbors Dean and Nancy, who put seed in their driveway for whatever avian wildlife found it first. Some mornings I'd drive off to work to find him sharing some sunflower seeds with the crows and scrub roosters. He was unfazed by most things, a calm monk among the loud chickens and car alarms also known as geese. He fit in. He lived well. He seemed happy. Living with a turkey so far had been pretty wonderful.

Many poultry-care guides advise against housing chickens and turkeys together in one coop. But what if the chicken coop in question only had one turkey? Disregarding convention, I decided to let TD shack up with the laying hens, two ducks, and my pair of geese. He seemed to have no qualms about the arrangement. At night the birds would fly up into their roosts on one side of the shed-cum-coop and the water-fowl and turkey would bed down in the straw on the other side. (I think they used to share one side, but the lower-level birds decided that waking up with chicken turds on their back wasn't such fun, so they switched sides.)

TD wasn't like the other birds, though. They were egg producers; he was an entrée. This tubby wanderer was my first-ever meat animal, and that wasn't an idea I took lightly. I had raised my own eggs, vegetables, and angora wool, but as a vegetarian. I had opted out of the bloodier side of home-steading and was generally happy I had. But my life on the farm was beginning to change me in important ways.

The moral vegetarianism I'd adhered to so strictly as a college kid was beginning to ebb. After holding the vegetarian line for so long, after reading all the books and handing out PETA tracts, it was starting to become less of my edible faith. I was finding it difficult to make an argument for not eating well-cared-for animals that were raised for food.

Raising food myself had changed my ideas about humans and animals, but not in that macho way, in which people think that other living things are theirs to exploit. No, this was not a question of possession. It was a question of equality. Back when I was lining my dorm-room desks with anti-meat stickers, I thought that equality meant protecting the rights of all animals to live without being dinner.

But folks, that's not how nature works.

Treating animals as equals doesn't mean treating them like people; it means *seeing humans as animals*. We are all pieces of one big puzzle. Before I became so aware of the life and death involved with everything we eat, I saw us as separate. Us and Them. But as I evolved from a consumer to a producer, I began to see humans as the animals we truly are. We're all food, organic matter that will either feed the soil or another animal. Eating meat is what predators do, and human beings are nothing if not predators. I had no problem with a wolf eating me or my eating a deer. We were all in this together.

I was losing my vegetarian religion, and because of it I was at peace with the fate of this particular mini-dinosaur. He would live out his days through the summer and into fall, and then somehow I would find someone to come to the farm and help me do him in and prepare him for the oven.

One weekday morning at Wayside, I was getting coffee and engaging in neighborly conversations at the round table (the wooden table, which once belonged to a famed fly fisherman, in the back of the country store that was the conversational hub of Sandgate). Sipping coffee and enjoying doughnuts and breakfast sandwiches were a collection of locals, mostly older men, who made sure to be at the round table early to weigh in on all matters town and country. One gentleman politely asked me what was new on my little farm. I was slowly growing a reputation as the girl who turned her rented cabin into a backyard farm and then wrote a book about it. I talked with the men at the table about the chickens, egg sales to coworkers, and the new sheep in their pen, and mentioned that I had this extra feed-store turkey that was getting huge and needed to be harvested for Thanksgiving. When I got to that last part, I must have spoken with a little extra pride. Raising meat is an honorable occupation in these hollows. The men beamed back, proud of this new country girl getting ready to enjoy her first-ever homegrown holiday meal.

Truthfully, I was shocked at how proud I sounded. Did this vegetarian really gloat that she had been raising an animal for food? I had, and without apology. I'd recently decided that TD would be my first nonvegetarian meal in almost eight years. It felt like the right choice. I'd signed up to farm meat, ensure that the life of this bird was good, and see that his death was as humane as possible. Also, I wanted to be a part of the big family dinners of my childhood again. I missed them. I missed swapping recipes with friends, splitting the wishbone with my sister, and sharing in the hard work of my father, who would spend the morning preparing the animal for the family table. I didn't want to turn up my nose at the holiday bird set before me.

Though I was getting more comfortable with the idea of returning to meat, I didn't feel ready to eat anything that had been raised in a factory. I was both concerned about the animals' welfare and scared of the diseases they carried; your average turkey in the grocery store might as well come with a biohazard sticker on it. But the fat turkey roaming in my backyard could lead the way back to carnivory. There would be no remorse for his death when I knew what his life had been like.

I didn't share any of this with the men at the round table. They had no idea how epic a meal it would be for me. I simply told them I was looking forward to the dinner, but I had to find a processor. Who would help me dress one turkey?

The men raised their eyebrows as I talked about the holiday bird, and Tom, a fit, clean-shaven man in his late forties, told me he would be helping the Pickerings process their birds and that I should join them. Pickering's Farm was a bit down the road, and they were longtime turkey growers. Locals had ordered their fifteen-pounders back in the spring, and now, as fall closed in, it was time to deliver on the deal. If I wanted to, I could bring TD down to the farm on the Saturday morning they were killing turkeys, and I could see to it that it was done outside in the open air on the farmers' land. I thanked Tom over and over and handed him my phone number. He said he would let me know when the big day was going to be.

I'd started this whole poult-to-processing project with mixed feelings. I'd purchased this turkey at the feed store because I knew that, no matter how I chose to eat, my family would be serving a bird for Thanksgiving dinner. Honestly, I think back on raising this meat as an act of kindness. If my family were intent on having a turkey, they could eat this beautiful, huge, healthy, free-ranging bird (for free, by the way) and not take part in eating an animal that had been overdosed with antibiotics, then strung up on an assembly line in some sunless slaughterhouse packed with turkeys, ten thousand to a room.

My bird was growing larger than I ever imagined he would. At five months he was no longer a cute little fluffball with scaly feet; he was a lumbering gobbler. Having reached his sexual maturity, he was also *very friendly* to the other poultry on the farm. And his romance didn't stop there. He was friendly with my grain bucket, T-post pounder, bags of chicken feed, and the pitchfork. Turkeys apparently don't raise the bar very high when it comes to getting some. This gusto was wearing all of us thin. And yet, despite his sexual faux pas and his mammoth size, he was a fine animal. The kind of bird Williams-Sonoma shoppers would pay more than a hundred dollars for. I was starting to look forward to the big day. I cleared everything out of the freezer.

The day for his slaughter was a cool, cloudy one in late October. I took Annie's dog crate out of the bedroom to line it with straw for TD's last road trip. We would be heading into West Arlington, just past Wayside, to Pickering's Farm. I trucked outside into the overcast morning and grabbed a scoop of scratch grains from the metal bin inside the coop. I called out my patented chicken call, "Hey, chickchickchick*chiiiiiikeeeeen*," which is as much of a hick cliché as is linguistically possible, and I saw the birds come a-running from all over the property. TD sauntered up as well, and as easy as picking up a watermelon, I lifted the turkey into the dog crate. If he had an idea what was about to happen, he didn't seem to be having any sort of existential issues about it. With my turkey loaded in the

Subaru, I headed down the mountain toward our date with the plucking machine.

As I came up to the farm, I saw Tom — and Pickering himself — down by the turkey pens. They had a basic but effective setup for their day's work: a suspended rope for hanging a turkey by the legs for the killing and bleeding out; a scalding tank for loosening the feathers; a giant plucking machine; a deep gulley of a pit for burying entrails; and a few steel tables laid out with knives, pliers, and hoses for the real grunt work.

I stepped out of the car and waved, a bit self-consciously (not sure how one should approach an abattoir). I shook hands with Tom and Pickering and thanked them for their help. We fell into the usual conversations people have in the country. Local news, weather, gossip heard at Wayside, the goings-on at other farms. I was deep into the chatter when a large steer wandered up to us. "Oh, him," said Pickering. "He's like a dog. Don't mind him."

I walked up to the big beef. I assumed he was a bottle-fed dairy calf, being raised here for the family's winter meat. Around here the dairies are lousy with male offspring, and you can get a beef steer for free if you know the right people and ask politely enough. To me, whose largest food animal to date was the one in the back hatch of my station wagon, he seemed beyond massive. All I could think of was "How could anyone eat all of that?"

I told you this farm was changing me.

When Tom and Pickering had time to deal with TD, they asked me to unload him from the car. When they saw me lift the giant bird, they were both bemused and impressed. The fact that I brought my bird to its slaughter appointment in a comfortable dog crate screamed "flatlander," but they sure couldn't scoff at what I had produced. My turkey was almost a third larger than the birds they were working on. Since he'd had an entire neighborhood to roam around and scavenge food from (on top of his daily rations), he was enormous. Tom whistled and complimented me on how white the feathers were, which was a new sight on this farm. The pen the farm birds lived in was well kept but hard-packed with dirt, which of course turned to mud in the misty rain we were sharing.

Tom carried my bird to the killing area. "He must weigh forty pounds, good Lord!" he said. I felt a hint of what being a real meat farmer must feel like. Grown Vermont men were patting this twenty-six-year-old kid on the back for raising a fine bird. Tom tied up TD by the feet; he didn't seem to fight it. As Tom went to grab his knife for the final act, I felt no need to brace myself. I trusted these neighbors and farmers completely.

In one swift motion of a sharp knife, Tom removed TD's head from his body and tossed it into the gut pit. I watched the dead head's mouth open and close while the body tossed around for a few final moments of firing synapses and squirting blood. It was in no way enjoyable to watch, but not upsetting. He had had a good life and a quick ending. Being instantly beheaded sure beats being ripped apart by a fox or run over by a truck.

In about ten minutes, TD went from a flapping white-and-red mess to basically what you see before your dinner goes into the oven. After he was scalded and defeathered, all that was left to do was to cut off the legs at the knee joints, pull out the viscera, and use some pliers to pull out the toughest of the tail feathers. Tom left the largest pinfeather on my bird's butt, so I would know exactly which of the carcasses was mine. To be of some use, I helped carry some recently prepared turkeys up to the farmhouse and picked over my bird. Mrs. Pickering, who did all the carcass detailing at the house, helped me look for stray feathers and other detritus from the butchering process. She was polite but seemed a little put off by the fact that I felt the need to use their staff to kill my own turkey instead of buying one of her birds.

Even though I was raising my own turkey, I was glad the Pickerings kept their small farm going, to help supply the area with local meat. I wanted to be a part of creating a sustainable food supply, and here in the more rugged parts of the Northeast, that meant meat and milk. There were plenty of local vegetables around, even a tofu factory up in the Northeast Kingdom, but in southwestern Vermont and eastern New York, dairy and meat were king. Within ten miles of my cabin, I could buy grass-fed beef, free-ranging pork, and pastured poultry. My local food economy was as much based on meat as it was on vegetables. If I wanted the small farms around me to survive, I needed to support them.

That power of the dollar was a huge reason I wanted to consider eating meat again. I felt that if I wanted to change how animals were being harvested for food, I'd better start voting with my paycheck and buying meat from responsible small farmers. Every dollar I spent at a local farm was another dollar that didn't go to support the cruelty of factory farms. As a Buddhist I still felt a heavy responsibility to avoid the suffering of sentient beings whenever possible. But I was beginning to understand that for me, abstaining from meat was less an act of kindness than it was a symbolic act that simply made me feel more Buddhist. It didn't in the least help the animals being raised in those horrid factories.

So I, the sole vegephile in the family, was okay with the fate of this particular turkey. I had raised it from a poult. I'd watched it live out its short life, and I was there to escort it to and assist in its slaughter. Now the bird, all twenty-eight pounds of it, was on ice in my freezer, awaiting the trip to my family's dinner table. I, the sappy liberal of the family, felt no guilt and suffered no emotional quandaries. So, of course, I assumed everyone else would be okay with it, too.

They were not.

When I told my family that I would be bringing home the turkey and would help roast it for Thanksgiving dinner, the other end of the phone line went slack. My mother was starting to feel uncomfortable with my homegrown bird — I think because she thought it would be unsanitary. Her assumption that my bird was not as healthy as a store-bought one is a testament to advertising, the sway of suburbia, and the huge distance from food you know by name. But in the end, my mother could set aside her concerns and nibble on TD. It was my sister who drew the line in the sand. She would not attend dinner if I brought to the table a bird she had met in person. This was not to be debated. It was the turkey or my sister.

I tried putting my foot down. I explained that I was bringing it and she would have to get over it. I was not going to waste this bird, and I told her I thought she was being ridiculous. She was in no way concerned about how Butterball turkeys were raised and had no problem eating them. To her these factory animals were so removed and so distant, a throng of meat mass she would never hear gobble, that eating them was less like eating meat and more like eating turkey-flavored protein globs. But I, I had *killed* an animal — an animal she'd seen strut past my cabin door a few months before. To her, this made her complicit in a murder.

Much arguing ensued. The family was split into pro-turkey and anti-turkey sides. My brother, father, and brother-in-law were for the farm bird; my sister and mother were adamantly for its staying in the North Country. My sister thought it was just plain creepy, and my mother cared less about the animal than she did about the possibility that her now far-flung family wouldn't be sharing a holiday. She didn't care if we all ate Chinese takeout, as long as we did it as a unit. My feed-store whim was about to separate us on a high Woginrich holiday. When I continued to try to persuade them to see the free-range light, my mom sighed into the phone and said, "You know what, Jenna? You win. We've decided to kill the dog and eat him for Thanksgiving dinner. Happy?"

In the end, I caved. I traded the giant frozen turkey for a sheepherding lesson. A week before I headed home to suburban Pennsylvania, I stopped at a sheep farm in Massachusetts to hand my trainer, Denise, the spoils of my summer. That Thanksgiving I would be eating Tofurky, again.

A NOTE ON THE DOOR

FARMING IS A BEAUTIFUL THING, but it takes a lot out of you. It's not so much the actual labor but rather the relentless responsibilities that do it. My humble little backyard farm was a constant commitment that rarely granted me the chance to travel, even for an overnight jaunt into Pennsylvania to see my family. Fortunately, whenever I really needed to be away for a weekend — for things like holidays and weddings — I could ask my neighbor Casey to look after the animals while I was MIA. He'd feed the sheep their hay, throw down scratch for the chickens, put pellets in the rabbit feeders, and make sure everyone had water. Having a neighbor who doesn't mind playing farmer and who lives within earshot of your farm is a handy resource. I felt lucky to have him. So when I left for Thanksgiving weekend with a pile of hay on the porch and a note on the door for him, I felt the animals would be well cared for. This was his fifth or sixth time watching over my menagerie, and the routine had become old hat for us both. I planned on surprising him with a gift certificate to Home Depot when I returned. He hated accepting payment for the farm chores, but I thought it was high time he knew how much I appreciated his effort.

After being away for three days, I was looking forward to being back on the farm. When I arrived home that Sunday, the sheep heckled, the goat nickered, the chickens clucked. All seemed well — at least until I got to the front door.

Pinned to the door was a typed letter and what appeared to be several pamphlets. I picked up the note, expecting it to be some detailed list of possible improvements or feeding regimens. Casey was really organized and had been known to install chicken coop doors and fix fences while I was gone. But this was no list. This was a very angry letter.

It started out with "I am disappointed and disgusted at the state of the animals and the property . . . " and went on to explain how horrible I had been to my livestock and to his friend's house. He indicated that changes would need to be made to the cabin and its backyard. No animal breeding, no additional building, and (he assumed, pretty bold for not being my landlord) all nonapproved animals would have to go. My heart pounded in my chest. This was a shock to me. In almost two and a half years of knowing me and watching the farm grow, not once had he come down on me like this. I noticed that Benjamin, my male Angora rabbit, was missing from the porch. Did he steal my rabbit? Was he holding him hostage? I read on. He accused me of neglect and abuse. The attached literature was on "proper" rabbit care and his letter explained that he had "rescued him." He said the yard was covered in feces, and the well water was probably poisoned, too. The letter explained how the animals had to go, and things had to change, that he had spoken to my landlord at length and I should expect a call. I started tearing up. Then I did the worst possible thing. I called him.

Like a beaten animal, I took his condemning words and even apologized. (Looking back, it was ridiculous that I admitted any sort of wrongdoing by apologizing, but I was upset and in "damage-control" mode.) I wanted him to understand why I did things around the farm the way I did. He said all the neighbors agreed with him but didn't want to say anything. Now I was really worked up. I was bawling into the phone, barely getting apologies and fears out of my mouth, terrified that I'd be run out of town and that the whole time I'd been falling in love with the cabin and with Sandgate, the whole community thought I was a monster. Logic flew out of my head. When I got off the phone, I called my landlord. I was still a mess.

"Did you talk to him?" my landlord asked, giving me the impression she wished I hadn't. "He has a tendency to *overreact* . . . listen, just listen. Jenna." And in what was to me the tone of a saint, she explained that she wasn't angry, nor did she think I abused animals, but the controversy was too much. Casey was an animal-rights activist, and after watching the animals this particularly rainy and cold weekend, he believed that the chicken manure, mud, wet animals, and caged rabbits were victims of a negligent part-time farmer.

My landlord was Casey's friend, and eventually, when she moved back to the cabin full time, she would be Casey's neighbor, so the last thing she wanted to do was anger him. She went on to explain that she'd never wanted the place to be a farm, that things had gotten out of hand. I started to recite the laundry list of green lights and okays I had gotten from both her and Casey. After all, her angry friend had helped me build sheep sheds and convert sheds to chicken coops. I wasn't thinking straight. Fear rolled into anger.

She told me to calm down but to start planning to change things a little. The goat, rabbits, and Saro's recent hatching of five goslings had to go. Those were animals I had taken in without her direct consent. All that was allowed to stay on the farm were the sheep, the two dogs in the original lease, and the chickens. I did the telephone equivalent of nodding, saying "Yes. Yes. Okay. Okay. Of course," trying to get the awful night of phone conversations over with.

She also asked me to vacate the cabin by spring (she was moving back for remodeling), and although I had a few months to find a new home for the dogs, sheep, chickens, and me, I would have to remove the fugitive animals immediately. She said she'd be in Sandgate the following weekend, and we could talk more in person if we needed to.

I would be ready.

I took a very long breath.

It took awhile to compose myself. When I did, I was officially livid. How dare that man steal my rabbit? Did he really think I was being abusive? The rabbit had a long coat that was matted in the back and his toenails needed clipping. He looked a little homely, sure, but I wasn't about to shear an outdoor rabbit's wool in winter on account of vanity. The mats could wait until spring. And while his cage wasn't winning any home and garden prizes, his hutch was still twice the size of what was required by state code.

I was no longer feeling like a girl with her tail tucked between her legs. I'd had property stolen, been accused of neglect, and been libeled in my new hometown. I felt I needed to defend my choices and my lifestyle. I didn't like that I was being perceived as a terrible farmer, and I *really* didn't like that this was the first time I was hearing about it. You'd think that if I were the talk of the town as a livestock abuser, someone would have subtly brought it up before I found it nailed to my front door.

As if the note weren't enough, thinking he was acting in the animals' best interest, my neighbor had "made improvements" that turned out to be downright dangerous. He'd lengthened the goat's tie-out, not knowing that I'd measured the chain to be just short enough to keep him from climbing over the rusty garden gate and hanging himself and also short enough to avoid the poisonous oak leaves in the opposite direction.

Everything I did at the farm had a reason. Long, matted coats were warmer than shorter combed ones. Old bedding under hooves might look ugly, but it kept livestock warm. Rabbits lived in cages to protect them from predators and keep their footprint small. The farm was just that — a farm. It was not a petting zoo or a set from Colonial Williamsburg. It was a place to produce food and animals for sale. Neighbors who had one dog or a couple of cats and didn't raise livestock got concerned and confused about the state of the place because it wasn't what they imagined a cottage farm to look like. There were no white picket fences or show ponies. My animals weren't living in a faded red barn like they do in your childhood memories. Rabbits were bred and sold, chickens were raised for eggs, roosters pooped on the lawn, and the sheep had walked the grass in their pen down to mud. It was never going to be on the cover of a magazine, but it was a real farm.

I sat down and typed an eight-page letter of explanation about how I farm. I would not have myself known as an abuser of the animals I had made a part of my life, my family. I would give a copy of the letter to any neighbor who wanted more information or thought I was doing something wrong. In phone calls with Casey and conversations with other neighbors (most wanted to stay out of it and had no comment), it became clear that they thought things like deep bedding the goat pen (a practice of laying fresh straw over soiled straw, to create heat and make compost when the older straw decomposes) was me being too lazy to clean the pen. I explained the length of the goat chain, the method of chicken bedding, refuted claims I was poisoning the well and destroying the property by homesteading on it. Rubbish, all of it!

I was proud of the little manifesto. It showed that I was well aware of their assumptions about me, but it also made clear that they were just that: assumptions. I hoped that by educating them, I would at least clear my name in the eyes of other animal lovers. That was the real damage control. I had no control over staying in the cabin; it wasn't my property. But I could save face.

That same neighbor who'd helped me build sheep sheds and mend fences was no longer the friendly man I remembered. He had lent me the rabbit cage but felt that it was too small for Benjamin now and wouldn't return it if he was going to live in it. It wasn't worth getting the police or anyone else involved because of the stolen rabbit. So when I e-mailed him that a new cage was on the porch and asked him to return my animal, he showed up with him a few hours later. Words were terse, and I said if he was really concerned about my animals, why not call animal control and have them assess the joint? I was joking, of course.

He wasn't. He called the police on me.

The following weekend, when Annie and I were returning home from taking a load of trash to the dump, I pulled back onto the private road that leads to the cabin and was surprised to see a large gray truck pulling out of my driveway. Ignoring it as a local who needed a place to turn around on the dead-end road, I drove past and parked the truck among the welcome clucks and honks of my poultry. In the rear-view mirror I could see that the gray truck had turned around again and was now pulling up beside me. A man in his late fifties, with an air of authority and a serious look on his face, jumped out and walked toward me. "Can I help you, sir?" I asked. Annie sat quietly by my side, tail wagging, on her leash. The man flashed his badge and introduced himself as a sheriff in Arlington, the next town over. He was here on animal-control duty.

You've got to be kidding, I thought. But I remained politely stoic.

He said he had spent the better part of an hour walking around the farm and checking on the animals. He said an anonymous complaint came in from a neighbor and then listed the litany of accusations: that I kept rabbits in tiny cages; that the property was littered with feces; that my sheep had no bedding; that my goat was in a cage that was too small (and also without proper bedding); and that I kept two huskies in cages all day.

He told me I had nothing to worry about. The animals, the farm, and the property were fine. The size of my rabbit cages was well above the minimum requirements of state law, the sheep had clean straw bedding, the goat was healthy and well kept, and my dog didn't look abused to him. I invited him inside to show him a clean house without cages or chains, and when I opened the front door, Jazz walked out to greet us. He petted Jazz, scratched Annie's ears, told me he was closing the complaint, and said he was sorry about the hassle. He shook his head, saying it was a shame to see places like this being called in to the police, that people who complain about a pile of chicken shit on a doorstep don't have any idea what horrible abuses he witnesses every day. He said if all animals on farms were as well kept as mine, he wouldn't even have a job as an animal-control officer. He said that most complaints about farm-animal welfare came from people with cats and dogs, not livestock, who didn't understand the difference between a goat pen and a dog kennel.

I felt all of my livestock deserved to be on grass and in sunlight, well fed, and kept safe from weather and predators, but this wasn't some sort of farm-beast day spa. These weren't pets, they were dinner. And if they weren't dinner, they were providers of future sweaters, or pack animals, or weed control, or garden pollinators. Before he left he gave me his home number and said to call if anyone complained about my farm again. He wished me a Merry Christmas and was on his way.

Moral victories notwithstanding, time was running out for me to figure out a plan for moving, and that scary fact was punctuated when I came home from work one day in early January to find in my mailbox a letter from my landlord. It told me that I would officially be asked to vacate the cabin and not to be alarmed at the certified letter that was coming my way. I knew this was coming. I had been warned by the landlord herself, over the phone.

Still, my stomach felt hollow, as if I were in a plane going through turbulence, or I'd accidentally missed the bottom step leaving the house. I stood outside in front of the Subaru's headlights to read the notice in the dark. It said — in a polite and calm tone — that she would be moving back to Vermont this summer to renovate to the cabin. She would soon be mailing me a notice to vacate on or before May 1.

I stood there, with that same feeling you get when you realize something you thought was real isn't, like when you finally understand your love for someone is unrequited (I'm actually an expert on this) or that Santa doesn't exist. That night, reading the letter, I finally understood that Cold Antler wasn't real, either — it was always someone else's.

I understood that four months' notice was generous, and ample time to pack up and move. I understood that a kind note sent to alert me of an eviction notice was courteous and friendly. My landlord was just doing what landlords do. I got that. I am not an irrational person.

I wish I could say something about how the note was actually some kind of amazing affirmation of my own plans and dreams and that the universe was coming together to make my life happen as I'd visualized it. But honestly, I didn't feel any of that. I was terrified. I. Was. Absolutely. *Terrified.* I felt as though I were sitting on a ticking time bomb and had no idea where I was going to end up, or how I was ever going to pull off getting my own farm. I'd always thought I would be the one sending a notice to my landlord. I thought this place would be mine for years, that I could live here until I was ready to move on to the next big thing and plan my life around that. But things had changed a lot since the holidays started.

If it had only been a matter of moving myself and the dogs, it wouldn't even be cause to blink. I'd done that before, when I was laid off from my job in Idaho. I had a new job and home lined up within a month, no sweat. But this was not just moving a girl and some huskies to an apartment in Bennington — this was trying to move an entire lifestyle. I had to find a place for me, a flock of sheep, a coop of birds, and two dogs in what amounted to only sixteen weeks. I needed to either get myself into a position to buy (and quick) or find another small plot of land that I could rent for another year while I saved. That second idea meant finding a landlord somewhere in the area who would welcome a small working farm. It wasn't impossible, but it was unlikely. If I couldn't buy a small place with a bit of land in time, I would have to find new homes for the remaining livestock and abandon the farm life for a while.

THE SEARCH BEGINS

APPARENTLY, I'M A SADIST, because even though I knew my credit score was a joke and my financial situation wouldn't allow me to even come close to owning a farm, I still scoured real-estate listings and drove around Sandgate scouting **FOR SALE** signs. I wanted to stay in Vermont (very much so), so I tried to find places in-state that were in my price range. Sadly, everything in decent shape with enough room to accommodate my small farm was well over two hundred thousand dollars. Out of the question. As a single woman with a modest income and no life savings to crib from, buying a place with a monthly mortgage payment the size of a new living room set was as far-fetched as deciding to take up the biathlon for the Olympics. I needed to find a decent place, cheap, and fast.

I expanded my search to within a thirty-mile radius of my office. It looked liked homes and land were cheaper just over the state line, in Washington County, New York. Online

I came across a little white farmhouse with over six acres, totally redone, and all set for farming with a barn and pasture. It was more reasonable than anything for sale in Sandgate but still fifty thousand dollars more than I could justify buying. I sighed, closed the browser window, and went outside to walk the dogs. I'd rented a farm in Idaho. I made this place work for a while in Vermont. If I had to, I'd do it again. There must be a place for us somewhere. Someone had to need money badly enough to rent to a woman with a menagerie of chickens and ovines, right?

The sad truth, though, was that I would probably have to find new homes for the poultry, the sheep, Finn, and the rabbits. I could find a dog-friendly apartment while I saved for my own farm. But it was crippling to think about how much time I would lose by taking a year off from the hoe and henhouse. A full season with chicks in the bathroom brooder or lettuce in the backyard was becoming as important to my living situation as indoor plumbing. I hated the idea of back-stepping like that, of missing the lessons and experience a year of gardens and livestock bestow on you. And I felt as though I had come so far: I finally had my flock. I had breeding Angoras, which were producing wonderful kits. I didn't want to undo all that. No, an apartment was a last resort.

In the meantime, word was spreading from Wayside that I was moving from the cabin, and those who were concerned about my situation had their ears to the ground. One freezing morning I was running late to the office, but my coffee addiction proved more powerful than my desire to be punctual. As I stepped inside the store, Nancy informed me that she knew of a place in town that might be available — a small cabin in East Sandgate that was owned by a family who had moved out West but still vacationed there from time to time. It had running water and electricity and was in decent shape, but it needed insulating and some serious winterizing if I was going to live there year-round. I got a lead for a contact, asked for the address, and decided to look into it. Rumor was holding steady that the old cabin was for sale, and in this economy an abandoned cabin might be exactly what I could afford. Nancy warned me that it was small, and only on one acre of land, but I could make that work for a few years or even longer if I had to. What I really wanted was some peace of mind, a place that was mine on paper. The only time I ever wanted to see paper nailed to my front door again was if I put it there myself.

The following week I took a long lunch break with my friend Steve to see the cabin during daylight hours. We got a little turned around but eventually stumbled upon it. I thought it was beautiful: a tiny log cabin with a stone fireplace and a loft, set into the side of a mountain, with a winding staircase that led to the road right next to the Green River. The roof looked sound, the windows were intact, and as far as my untrained eye could ascertain, it looked like a solid house. Steve nodded in agreement. It wasn't fancy, that was for sure, but it was cheap. Someone told me they were asking sixty to seventy-five thousand dollars for it and the land, and that was a mortgage I could afford. We drove back to work chatting and laughing. I was filled with hope and Steve was filled with the Italian grinder from Wayside he'd picked up for lunch. We're both made happy fairly easily.

The small cabin was starting to seem like the ideal solution. I had spoken with the neighbors about the property, and they seemed thrilled at the idea of sheep ruminating outside their windows. One potential neighbor had offered me the use of his barns and pastures in exchange for having my flock mow his lawn. Suddenly, my one acre was now next to a rentable pasture with more land than I knew what to do with. I got weak in the knees at the thought of starting my own lamb and wool operation right here in town, not to mention being able to sit with a banjo on the porch of my own cabin. It could be heaven.

The family who owned the little cottage seemed genuinely wonderful. We talked for hours about my life, my goals for the place, the property, and my lifestyle. They explained the importance of the cabin to their family and how much nostalgia came with the land. They said they had named it the Foothold, since it was where they felt the most grounded in their lives. I was so touched I had to sit down. I explained about the livestock and bees, the gardens and dogs. I wanted them to understand that this wouldn't be a deer camp or a second home to be used only for ski vacations. I would do the place justice, make it my home. The matriarch of the family (who held the deed) seemed smitten with the idea of a young gardener with a good sense of the town taking the keys.

Hoping to win them over, I sent postcards and a package, explained in letters how much I wanted to stay in Sandgate, and even collected a few letters of recommendation from the locals. My attempts were well received, but we never seemed to get to the nitty-gritty of deal making.

Finally, one night I explained rather bluntly that I was out of a home in a few months and had a whole pack of farm animals that would be homeless, too. I didn't have the means to secure a mortgage and was curious if they'd consider a rent-to-own option or holding the papers (meaning I would own the mortgage but they'd hold the deed until a certain amount of money had been paid). Both would secure me legal rights to the place without having to go through the agony of the housing process with banks and real-estate agencies. The family said they'd consider it after talking with their lawyer. I held my breath. I wanted the Foothold to work out.

It didn't.

In the end, they simply wanted too much money and couldn't cover the cost of making the camp into an all-season home. It was an odd situation to find myself in. Because they would not agree to owner financing or renting-to-own or anything else that might land me inside the heavy wooden doors, it meant I had to get approved for a mortgage. But no bank would approve a loan for a first-time homebuyer who was purchasing a shack without heat or insulation. Since I didn't have the cash or a bank behind me to offer them the full price, I had to decline. They wanted cash on the barrel-head, and I had $586 in my savings account. So I lost my footing on the Foothold.

Stress was high. With the fugitive animals in foster care — Finn had been picked up by a family in Bennington, the rabbits and goslings were staying with a family a few towns away — I could at least stop worrying about their welfare while I figured out my new address. But even with a few of the animals in farm purgatory, I still had twenty chickens, five hundred pounds of sheep, and two hairy dogs to move. Finding a rental was seeming less and less realistic. The ideal situation would be to buy a place. My own farm. And I was running out of time to figure out how.

So I did what any clueless potential homebuyer does: I kept searching. I found a place online and called the number on the listing. It was a 1700s farmhouse with a woodstove and five acres — enough land to keep me busy for quite some time. The price was a little high, but in the housing market we were in, it seemed reasonable to offer less. I left a message explaining my situation, briefly, and the MLS listing number.

A few days later a perky woman (suspiciously perky) called me back. She sounded thrilled at the idea of my owning a farm and wanted to help me out and asked me a hundred questions about my financial situation.

"You make how much a year? You have how much saved? Your credit score is what?"

None of my answers made her tone lighten. She calmly explained I would need a down payment of somewhere in the neighborhood of twelve thousand dollars for a home under a hundred and fifty thousand. That was a crippling amount of money to me — almost a third of my salary. To make it appear overnight would be impossible.

I realized that if I was ever going to have my own farm, I needed to start squirreling away money. I started saving. I didn't buy coffee at Wayside. I ate spaghetti for dinner for weeks. I sold my banjo (again), some artwork, antiques, and music online, and ended up a month later with two thousand dollars in my new savings account, barely enough to cover a home inspection, a good-faith deposit, and a moving van. Things were starting to look hopeless. I worried that I'd be posting on Craigslist soon, giving away chickens, geese, and sheep to good homes. I didn't sleep well.

Then one day everything changed. The price of a property in Washington County dropped by more than fifty-thousand dollars. It was in a town called Jackson, just over the state line in New York. It had everything: a white farmhouse, a pasture, a barn, a chicken coop, an artesian well, a stream, and a pond on six and a half acres. I did some math and realized I could afford the mortgage for roughly the same cost as my old rent and car payment on the Subaru (which I had just paid off). This could, just maybe, be mine.

I called the agent and asked if I could see it. He said of course and that the sellers were highly motivated. A retired couple had restored the place over the past four years but now wanted to move south and be with their family. He explained that the heating system, well, insulation, electric — everything — had been redone. The house was in amazing shape, but it was hard to find the right buyer. It was small (roughly eleven hundred square feet) and had warped floors and low ceilings. It had only two bedrooms and one small bathroom downstairs. It was built around the time of the Civil War and still had the old tree-trunk rafters in the attic. It sounded wonderful. And to a girl living in a shack, it was a mansion.

I drove to see it the following weekend. Leon, the agent, walked me around the property and told me, almost apologetically, that he wished it was summer so I could see the place in its full glory. But I could easily imagine the trees flush with leaves and picture the giant maple in the front lawn bursting red with fall foliage. It was remarkable. Inside the small house, things were quaint, albeit a little wonky. The floor was old and uneven, the rafters in the living room were exposed (which I loved), and the weird cave of basements and the attics and mudrooms behind the house seemed cobbled together almost as an afterthought. It was funky, sure, but it had character.

I told Leon I loved it, wanted to know what to do next. I explained my limited savings but decent employment history and a perfect rental record. He told me to talk to his mortgage broker, Jim, who knew a lot about financing homes in Washington County through, of all places, the USDA. Leon explained that the Rural Housing Development Program let single-family homebuyers get a property with no money down (my ears pricked) and still qualified me for all the first-time homebuyer incentives the government was pimping to resuscitate the market (ears fully pricked). After the tour, and that talk, the idea of actually owning the small farm trotted from the world of fat chance into the realm of possibility. Could there be a perfect storm of luck, a recession, no-money-down farm loans, and desperate sellers? Perhaps.

After the first couple of "near purchases," I told myself that I should know better than to get emotionally invested in this house. I should be solid steel during meetings with the agent. I should care about it as much as I care about the filing cabinet in the office. Poker face. Stiff upper lip. Walk away like a champ. But getting emotionally invested in my lifestyle was what had gotten me where I was. It was what had driven me to move cross-country (twice), start a renter's homestead, write a blog, author books, and try to change my patterns of consumption. So really, I'd been attached to the house before I knew it existed, before I saw it. It was the embodiment of a life I craved deeply. I was sure my claw marks were already on the deed.

I wanted to go home. I wanted it so much, my ribs rattled.

Then I found out another person was being shown the farmhouse. Hearing that was an unexpected punch in the kidneys. I had already made my offer and the owners had countered. We eventually came to an agreement that made them happy and cost me only fifteen dollars more a month. I wasn't out of the woods yet, though. The other people could still beat my offer. And I still needed to be approved for a mortgage, which was the razor's edge of this whole thing. Since I had dedicated myself to fixing my credit, I'd raised it twenty points, but it was still twenty points below what the lenders wanted. If they decided against me, I'd be out the dream. And I hated the thought of the property going to people who wouldn't use the land.

It was ridiculously stressful. I knew in my logical mind that there were other homes out there, and I am one of those people who believe that everything happens for a reason . . . but to lose out on this place, at this price, near my work, with the threat of eviction hanging over my head, would be too much.

GOING HOME

THE ANIMALS WERE NOT CONCERNED. As I was being emotionally accosted by the stress of my pending eviction while trying to pull off a Hail Mary mortgage, farm life continued as it always had. The chickens kept laying eggs and pooping on the porch. The dogs slept and played as usual. My three sheep (Sal and Maude remained, but I'd returned Marvin to his previous owners, at their request, because they missed him; I had replaced him with a little black lamb I named Joseph) were as normal as rain. They ate, *baa*ed, and shuffled around their pen and small pasture as always. There was some comfort in their apathy; I interpreted it to be confidence that all would be well. It was delusional, of course. I knew their lack of anxiety had to do with the comfort of their routines and not their faith in me. But routines also brought me comfort. Feeding buckets of grain, refilling water fonts, buying a few bales of hay from Nelson . . . it was normal work. And no matter how much my life swirled around the possibilities of locations, that work stayed the same. I cherished this.

I was e-mailing my mortgage broker every day, trying to get some sort of update on the possibility of qualifying for the note. I had compiled all the paperwork, paid off my credit cards, and boosted my savings as much as possible; but none of this mattered if I couldn't get my credit score up to the magical number the bank wanted. I checked my score constantly online and prayed that the fifty-point jump I needed would occur. Because scores take awhile to reflect the paid-off bills, my broker wanted me to wait until the last possible minute to officially apply. It was like waking up and walking on glass every day.

To make myself feel better, I tried to line up some plan-B housing. I found a guesthouse I could rent in nearby Sunderland, Vermont. I heard word of possible foster homes for my flock until I found a place where we could all live together again. I didn't allow the idea of living without my dogs enter my head. That was the deal breaker. If I had to find homes for the farm animals, I would. But I would not give up Jazz and Annie under any circumstances. We had lived together since my first postcollege job in Knoxville and had traveled across the country together twice. We wouldn't be separated now.

At some point, though, I realized I'd already made the decision in my heart that I would be buying a farm. That's all there was to it. This was no longer something that I wanted to happen: It was something that was going to happen. Years of homesteading on rented land were starting to wear me down and were causing more anxiety than peace. I needed to be in a place of my own, a place where I didn't have to worry about violating my landlord's rules by hatching goslings or about whether or not I could expand the garden to accommodate more pumpkins. I was living a life of increasing self-sufficiency on a small homestead, but it was on someone else's land. The irony was becoming heavy. I realized that I could not waver. I couldn't give even the slightest consideration to the notion that the farm would not come through. I was in the last two months of waiting and planning. My official day to be out of the cabin was May 1, and March was halfway over. There were a real-estate agent, a mortgage broker, home sellers, a frustrated landlord, and a crazy neighbor all banking on me.

I decided that if the mortgage didn't go through, I would find some other way to buy this house. Already the broker and I were discussing options. The USDA standards were a bit higher than those of the Federal Housing Authority, and there were many hoops to jump through. The standard government-backed loan was administered by the FHA. It was a little more flexible with credit scores, locations, and projected limitations. The USDA's program had a lot more restrictions but was worth it for the cash I would save. If I needed to take that route, I would. I was also ready to talk to credit unions, other banks, private lenders, or donors or consider possible rent-to-own or note-holding scenarios. The bit was so tight in my mouth, I was grinding away at my own teeth. As the weeks fell into the longest part of winter, the idea of owning a farm became a certainty. My mind was made up. It was that simple.

Then one afternoon I sat down with a hot cup of coffee and wrote on a sheet of paper everything I wanted, sketching between the words. I wrote about a little hillside and a strong farmhouse. Behind the words were drawings of sheep dotted along a steep pasture. I wrote about a day on the farm in October, smelling the apples in the trees, as we watched the now fleeced-out spring lambs gamboling toward us for evening grain. I say "we" and "us" because I drew myself with a dog on the hillside. This dog and I were partners, watching the flock. I kept writing about the Jackson farm and what it would be to me, how I would turn it into a proper sheep farm and raise lamb and wool and work beside my dogs. I was getting obsessed with the spell I was casting at my kitchen table, and the coffee I had set out to sip was now cold. Mentally exhausted and emotionally excited, I set down the paper and pencil and leaned back in my chair. I let out a sigh and looked at my scribbling.

Hmm.

I recognized that farm; it was the one I wanted to buy. I recognized my truck, the sheep, the topography of the land, but I did not know that black dog. If a working border collie was integral to the manifestation of this life, where was he?

I knew the answer to that. He wasn't here because of how awful my experience with Sarah had been. Also, I knew I couldn't raise a third dog on my property without the crazy neighbor telling my landlord. I had stayed a member of NEBCA since the first sheepdog trial I'd gone to, summers before, but I figured that having my own working pup was a life away. Well, was that not what I was crafting this very morning? Was I not in the process of talking to banks, moving, buying a farm? Had I not made up my mind that this would happen?

I would be a shepherd. It was time to get a sheepdog.

I started looking for a pup that same day. I knew what I wanted: a rough-coated, large, male working dog. A classic black-and-white, blocky sheepdog with dark eyes and strong instincts. Around New England, this dog seemed rare. So many of the serious trainers and breeders had smaller working dogs or ones with smooth coats, like the ones Highland shepherds in Scotland seemed to win big trials with. I contacted every local breeder in the NEBCA directory, but all of them had pups spoken for, or no litters planned, or not the kind of dog I was looking for. There was a rumor that two dogs I had drooled over at trials — a dog named Tot and a bitch named Song — had had a litter, but all of them had homes by now.

Eventually, I found a border collie breeder in Idaho named Patrick Shanahan. His outfit was called Red Top Kennels. I was impressed by his trialing experience, and his dogs looked exactly like the beasts of my dreams. One in particular, a handsome dog named Riggs, was photographed with him at some big trial in Europe and had won top honors. As I clicked through, I saw that the man had litters available a few times a year. His rates for pups were hundreds of dollars below what they cost here. It looked as if a fine pedigree from working lines would cost the equivalent of two car payments. I could do this. Hell, I'd sell my fiddle if I had to. I sent him an e-mail explaining my goals as a beginning sheep farmer and that I was interested in a puppy. He had a litter due in May sired by Riggs out of a bitch named Vangie. Both were rough coats and good workers. If I was looking for a farm dog, he could ship me one. I mailed him a deposit for a black-and-white male, if one happened to be born.

He e-mailed a few weeks later that three healthy pups were out of Vangie, and two of them were males. He sent a photo of the little Oreo cookie blobs, and I melted. In a few months, one of those boys would be in my arms. In a year, he'd be by my side at herding lessons, learning this life with me. I looked at the pups and told him I was drawn to the boy with the black spot on his back. Patrick kindly reminded me that my pup would be the one he felt would suit my needs as a farm dog, not the one I preferred based on looks.

But my interest in that pup had nothing to do with the spot; it was in his eyes. Even in the photo, the little four-week-old seemed to be gazing right at me. He was already my dog, twenty-eight hundred miles away. I wanted to name him Gibson, after the guitar company. I'd always dreamed of owning a vintage 1950s J-45 acoustic guitar, but the price and rarity made it seem impossible to think that I'd ever own one. Yet here was this dog, amazing in his own right and part of a dream larger than any old guitar. The Gibson J-45 is deep black with a tobacco sunburst across the top. It has hints of brown and white and the ability to sound gentle and bright or strong and imposing, depending on the situation. I wanted a sheepdog with that same ability, one that could be calm around lambs, tough around rams, and grounded in his work and life.

His name was set in my heart. Gibson and Jenna: partners. My pup was on his way, and the farm was one step closer to being real fur and soil in my palms.

After all that stress and worry, everything worked out in the end. I got the call at my office that the mortgage had been approved!

With backing from the USDA's Rural Development Program, I would have the money I needed to buy my farm. And the good news didn't stop there. Thanks to a seller's concession earlier in the process, I would have enough cash to take care of the closing costs, lawyers, inspections, and movers, with some left over to float me until my next paycheck. It wasn't ideal, but it was enough. It was more than enough.

Just hours before the bank application was due, my broker ran my credit score. It wasn't shining, but the months of paying off bills, scrimping, and saving was enough to get me twenty points above the waterline.

I collapsed back into my chair and felt such relief, I somehow lost the ability to cry or smile. I was still shocked. People say "shocked" all the time in casual conversation; usually, it means the person was surprised or rattled in some way. I was actually, physically, in some sort of suspended animation. I sat there in my desk chair, some Photoshop document open on the screen, barely noticing the people working and chatting around me. With one simple phone call, my life had changed forever. I went from being a panicked, tail-tucked renter on the verge of losing everything to an actual farm owner, a genuine freeholder. I would soon hold the deed to six and a half acres of pasture and forest, a little white house, a stream, a barn, a pond, and some scrappy outbuildings.

The closing would be in two weeks, my broker said. I'd sign the papers, then the previous owners would hand me the keys and leave. This all raced through my mind for several minutes as I sat there. Not working. Not talking. Just staring at my screen in the awed concoction of relief and gratitude. Finally, a smile started to add color to the empty canvas of my expression. I started to grin like an idiot.

I was going home.

As the closing date on the house grew near, I realized I was starting to feel off. Something was wrong, but I couldn't put my finger on it. I started to lose my appetite; sleep was harder to come by. Thinking it was the nerves of a first-time homebuyer, I wrote it off. What else could it be? For years I had been slowly crawling uphill. I had fallen in love with farming, and fallen hard. I had a romance with homesteading in Idaho, my honeymoon in Vermont, and now I was less than a week away from sleeping under the roof of my own farmhouse, a place where my bedroom view would include my own barn and land. There was a puppy waiting for me in Idaho. I had more money saved, more bills paid, and the highest credit scored I'd ever had. And yet, as the day came to signing that mortgage and getting my title, I felt drained.

I didn't let myself linger on it much. I pushed aside the odd feelings because they confused me. I couldn't piece it together. What had happened in my mind to take the joy out of this adventure? Then the confusion over the thoughts in my head started to make me feel guilty. How dare I feel anything but happiness? And so I went about the business of packing up the cabin, making arrangements with the local moving company, returning items I'd borrowed from neighbors, and planning how I was going to move the farm animals. I started jogging to take my mind off the turbulence in my head. I didn't feel like eating. I lost five pounds in a week, although I didn't notice until someone complimented me. I was running on fumes.

I realized, slowly, and in many shallow cuts, what was bothering me. There was no orchestral revelation, no falling to my knees in tears. But I was slowly coming to understand the heaviness I'd been carrying around. I was about to turn the key on everything I'd ever dreamed of . . . and I had no one to share it with. I was moving into my first house and starting my own farm, alone. It crushed me.

It wasn't that I thought I needed to have a man in my life to feel complete. Nor did I believe that I needed a partner to start the farm (well, at least one that wasn't a sheepdog). But while I found such joy in making the farm happen and watching the process unfold, I still wanted to hug someone when I got the call about the mortgage. I still wanted someone to kiss me on the forehead and tell me he was proud of me. And I still wanted him to grab a pitchfork and his own dog's lead and walk out to the barn with me for morning chores. I missed someone I'd never even met.

The funny thing is, I never thought about loneliness when I was renting. I think there's something naturally impermanent about a lease. I had signed many of them, in four states now, and not once did I get the sense that it was a long-term decision. No, to me signing a lease was like making a contract with myself that I wasn't yet dedicated to anything. As soon as the lease was up, I would be free: free to move back to Pennsylvania with my family, or to Boulder or Austin, or to God knows where.

But signing a mortgage was a serious commitment. It was a marriage with a place, and it said to the world that I would be staying put. I'd be living permanently in one place, following one sacred goal, far away from the place where I grew up; going back would be harder than ever. A lot of people do this, and some are lucky to do it with a spouse. I wasn't willing to wait for him to show up to start lambing on my own pasture, but that didn't mean I wasn't heartsick at the thought of going it alone. After all, soon I would be giving this farm all of myself. I'd be watching over and caring for dozens of animals, gardens, and bees. Who was going to watch over me? Suddenly, Washington County became a very scary place.

People often told me they thought it was so brave of me to move to Tennessee, or Idaho, or Vermont by myself. I would listen to this and say "Thank you," but the entire time they were praising me, I wanted to shake them and tell them, "*Don't you see the real story here? I'm not brave at all. I'm just terrified of regret.*" I started a new life three times in five years, but I was never smiling as the planes took off. It was always with fear deep inside me that if I stayed put, I would be missing out on whatever lay ahead in that foreign world. This was true about Tennessee, the place that changed me into the person I am today. And it was also true of Idaho, where I dipped my toes into homesteading and fell in love with it. And it was certainly the case with Vermont: a place I had come to see as home for the first time since I was in my parents' house. Now I was moving once again to a new state (my fifth in five years), and even though it was only nine miles from the Sandgate cabin, it was an entirely new place with a completely different vibe. I prayed it would be kind to me.

The first thing I thought of when I woke up on closing day was crows. I have a personal superstition that spotting crows in pairs is an auspicious sign. For me, two birds — side by side in flight, perched in a tree, or hopping along the side of the road — is an omen of hope. I have no logical explanation for this; it simply feels correct. When I tried to figure it out, I found in my research that crows are seen in groups only when they're young. So what I had been smiling at was actually an interdependence I didn't understand but deeply appreciated. A pair of crows is a sign of necessity, teamwork, survival, and, thus, hope. It's how the young animals find their footing in the world. The gypsy in me needed to see those birds before closing on the Jackson farm — a mandatory blessing. As I rolled deeper into the quilts to try to gain a few more minutes of sleep, I silently prayed that a pair would find me before pen touched paper. I took a deep breath and got up.

I put the coffee on the stove and placed the leftover quiche in the oven. As the percolator rocked and the cabin filled with the smells of the warming breakfast, I invited Jazz to join me on the couch. He leapt up and buried his head in my chest. His tail thumped as I held his face and kissed his forehead. It was far too early and too dark to do chores, so while we waited for daylight to catch up with us, I petted my dog, then grabbed my old guitar and played a song with breakfast.

By the time we'd had our fill (I ate a little, too nervous to enjoy the food) and I had enough coffee to scare normal people into caffeine celibacy, it was time to get outside. The last day of chores as a Vermonter.

It was the swan song for that incarnation of Cold Antler. In a few weeks the cabin would be empty, the yard quiet. Not a single rooster would break into morning yodels here. I didn't know if the neighbors would be heartsick or relieved at the change. Truthfully, I tried not to think about it. While I went about the morning rounds of sheep, chicken, and rabbit care, I listened to soft music through my headphones and let myself lose focus. Soon, though, I discovered just how hard it is to meditate when you realize you've acquired thirteen more new animals overnight.

The big speckled doe had given birth to a giant litter of kits. Inside the nest box were a baker's dozen of wriggling little babes — some pink, some spotted, some near-black. They were all alive and well, and their mother was doing fine. This marked the first litter of meat rabbits at the farm, and the fact that it was on the day I closed on my new home seemed almost worthy of a script. I reached into the furry nest box and pulled out a tiny kit. I held the newborn in my palm, warm and close. I watched the tendrils of steam come off its fragile, chubby body in my hand, then quickly returned it to its family. I smiled at the small success. Even if crows didn't come, kits had.

With the blessing of the new litter, I headed inside to prepare for the big day. I had originally planned to bring Jazz with me, but when I realized how much driving was involved — and how long the day might be for an elderly animal — I decided to let him and Annie rest at the cabin. But the idea of closing alone was depressing. I wanted someone or something with me to share in the sublime moment. I grabbed my beat-up guitar and set it by the front door: That would do.

The old guitar had become a good friend. If a dog could not join me on this fine day, this scrappy guitar would be a fine second choice. I loaded her in the truck and turned on the music, and we drove down the mountain toward the rest of the day.

It was all starting now.

On the way, I pulled into the Wayside to grab a cup of coffee. I also wanted to pick up something small to walk into the new house with. The store had been my home-away-from-home since I moved from Idaho, and I wanted a piece of it to join me. I found a glass case near the magazines with a pile of tiny, jadelike Buddhas in every shape and size. I found a small Japanese Buddha (not the fat, happy one holding coins) and bought it for a few dollars. As I climbed back into the truck, I set the little green statue on the passenger seat. Together, the Buddha, the guitar, and I drove to the bank. We had not seen a single crow since leaving the cabin half an hour ago. I gripped the steering wheel tighter.

After the cashier's check was made and my obligations met, I headed over to Chem Clean to service the truck. I pulled up next to the air nozzle, and Chris, a neighbor and the attendant, noticed the case in the front seat. Chris is a fine guitarist, so he helped himself to a few tunes while I checked the tire pressure. As I pressed air, I could hear "Black Bird" playing from the other side of the Ford. It was beautiful. I stuck around for a while to talk with Allan and Suzanne (who run the shop) and got to meet their two new Siberian husky puppies. I held the little girl pup in my arms while her brother ran around in circles on his blue lead. Not many gas stations in America offer a concert and puppies.

I drove faster than I should have. The music was loud and emotional. I was listening to a new passion of mine, Gregory Alan Isakov. I was drumming with my thumbs on the wheel while the violins started to shudder in "The Empty Northern Hemisphere." As I crossed the state line into New York, I felt the rush of quiet panic mixed with the excitement of something new. A few miles up the road, as the song galloped into the bridge, a pair of crows flew over the truck. I let out a long exhale. Everything would be okay.

The rest of the afternoon was a blur. A pile of meetings, lawyers' offices, paperwork, and handshakes. The whole time I was signing documents, I couldn't believe this was actually happening. The idea of getting my own farm just a few months ago was a fantasy. My credit score was horrible. I didn't have a savings account. I had no real plan to find or buy land, and yet here I was, four short months later, being asked if I wanted extra title insurance and being handed a set of keys. When all was said and done, I stood up from the heavy wooden desk and realized I was shaking, like I was in love.

I think I was.

I drove back to the Jackson house, my house, as it started to rain. I had a thick packet of papers and a smile that would not hide. I kept checking my phone to hear word from friends and family. Two coworkers would be coming over with pizza and beer later to celebrate. My parents showered me with congratulations. I thanked them all, over and over. But despite their kindness, I could not wait to hang up and go *home.* I wanted to walk around the property, planning animal housing, running extension cords, making the place come back to life. That dead farm was about to get a few hits of Jenna. It would resuscitate, and thrive, and feed people again. I was drunk on the dream's turning into reality. I wanted more. I wanted to be in the house. I drove like it.

Then I almost hit Stumpy.

Stumpy was an aging golden retriever who happened to be strolling down the middle of Route 22 (a busy rural highway), not even trying to dodge the passing cars. People who noticed him cut him a wide berth, and others slammed on the brakes. I knew his name was Stumpy because I pulled over and hollered "Hey! Dog!" and he trotted toward me. A line of cars was slowing to watch this dumb girl try to flag down a senile dog, but I ignored them. (If that were my dog, I'd want someone to call the name on the collar.) So, Stumpy came to me and sat with me on the side of the highway, and we got acquainted. I called his owners, and they said they'd be down to pick him up. While we waited for them, I told Stumpy about my day. I was grateful to have met him; he got me to slow the hell down and just sit. His horrible pedestrian ways let me take in what was actually happening. And I was secretly happy to have an arm around a dog. Dogs are my people. We talked like old friends. I was somewhat sad to see him hop into his owners' car.

Keys in hand, dog rescued, and barely a mile from my new home, I drove a little slower. The road that led to my new address was a winding one, curving around new-growth forests that were once pasture. I pulled into the driveway, parked, and grabbed the Buddha and the guitar. I opened the door and stepped into the warm house, which was filled with rays of afternoon light. It was so much brighter than the cabin was, even on its best days. I set down my keys and the guitar case and walked around, trying to catch my breath. I felt the stairs like they were lightly vibrating with electricity. I walked the rooms as if the walls were lined with paintings. I drank it in. Everything made me feel brand-new.

I sat down on the floor, opened the guitar case, and played a song. "Upward Over the Mountain" rang through the old farmhouse and echoed upstairs. I played it like it was the last song I'd ever get to play.

Mother, don't worry; I've got a coat and some friends on the corner.
Mother, don't worry; he'll have a garden. We'll plant it together.
Mother, remember the night that the dog had her pups in the pantry?
Blood on the floor and fleas on their paws and you cried 'til the morning?

So may the sunrise bring hope where it once was forgotten.
Sons are like birds flying always over the mountain.

Sitting there, sweaty and excited, daunted and alone, I sang. I was overwhelmed and happy. Really happy. But understanding and feeling those things all at once, I started to cry. It wasn't a cry that belonged to any particular emotion. It was a homily and a eulogy; hope and fear; desire and despair. I just cried. I held a black guitar against my chest, shook, and cried. Some things can't be helped.

So much of my story is about wanting. To finally have it is a relief so complicated and beautiful it breaks me to understand it.

Crows fly. Buddha sits. I farm.